WHAT AI REVEALS

How Artificial Intelligence Exposed What Was Already Broken, and What We Can Build Instead

Bob Hutchins

First published 2026

Contents

Preface

I used AI while writing this book. That fact belongs near the front, because the book is about the line between assistance and substitution.

The tools helped me research sources, test outlines, generate counterarguments, and notice weak spots in my reasoning. What the book means, which examples come from my life, and what I would defend in public: those decisions stayed with me.

My process was uneven, as real writing usually is. Some days I used AI heavily for bibliographic work and argument testing; other days I closed the tool entirely because the writing needed silence, friction, and a notebook. Several chapters changed shape because the act of drafting showed me that I did not yet believe what I was saying.

AI can make a draft arrive faster. It cannot give you the understanding that comes from staying with a difficult thought until it becomes your own. The understanding is in the wrestling. Remove the wrestling everywhere, and the work may look finished while the person doing it remains unchanged.

This book is mine in the sense that matters. The judgment is mine. The voice is mine. The responsibility is mine. Where the tools helped, I have tried to say so plainly. Claims that depend on research I have checked against the sources, and where the argument goes beyond the evidence I have tried to make that visible.

Writing the book also forced me to live inside its central test. I had to decide which parts of the work could be assisted without hollowing out the work itself, and although that line moved by chapter, the standard did not. If I could not explain it, defend it, or stand behind it, it did not belong under my name.

Show up. Struggle. Stand behind it.

I wrote this book with that standard in mind. Read it the same way, as an invitation to test what still deserves your attention in a time when so much can be generated on command.

A Note on Sources and Method

This book draws on 3 bodies of evidence.

Empirical research includes peer-reviewed studies, institutional reports, public datasets, and published analysis from education, work, health care, creativity, media, and AI governance. Where the claims concern current AI capabilities, labor markets, regulation, or environmental impacts, the research is recent. Where the sources clarify enduring human concerns, the book reaches for older work in media ecology, organizational psychology, philosophy, education, creativity, and contemplative studies.

Case material appears in 4 forms: publicly documented cases such as Clarkesworld, the Steven Schwartz brief, and the xAI Memphis controversy; anonymized client or practitioner examples; composites drawn from recurring patterns in my professional work; and explicitly fictional portraits used to illustrate a documented pattern. I identify these categories in the text where the distinction matters, especially in chapter 9.

Interpretation is my own account of what the evidence reveals about human judgment, trust, presence, and responsibility. Research can show patterns, limits, and correlations. The argument about what those patterns mean is mine, and I have tried to keep that distinction visible so the reader can judge the evidence and the interpretation separately.

The AI terrain moves quickly, and specific capability claims may age. The deeper argument should not depend on a single model, benchmark, or company announcement. The book asks what becomes valuable when execution becomes cheap.

The goal is not to make AI feel simple. The goal is to make the human stakes harder to miss.

INTRODUCTION

The Mirror We Didn't Ask For

In February 2023, Neil Clarke closed the submissions portal at Clarkesworld Magazine.

Writers had not stopped sending stories. He closed it because of what was flooding in: within weeks, the magazine was receiving hundreds of pieces that looked oddly alike. The sentences moved cleanly, the structures held, the dialogue sounded plausible, and still something was missing. The stories had the surface features of fiction without the strange pressure of a particular life behind them.

That moment clarified something larger than a slush-pile problem. ChatGPT had crossed into public life at the same time, with a UBS estimate reported by Reuters placing it at 100 million monthly active users within 2 months. People were startled that a machine could produce so much text so quickly. The more uncomfortable discovery was that so much human work had already been trained to look like text a machine could produce.

AI did not create the hollowness. It exposed it.

Schools had been rewarding polished compliance. Workplaces paid for documents that proved effort more than they changed decisions, media systems wanted volume, and the creative platforms kept ordering the familiar. The machine learned from the world we built, then reflected that world back to us.

I have spent the last several years inside this terrain, training educators to think carefully about AI in classrooms and advising organizations working through responsible adoption. My research has examined what generative AI does to creative authorship and authenticity. The same pattern keeps appearing: the real issue is rarely the tool by itself but what the tool reveals about the systems around it.

We have been here before, in smaller ways. When the Remington/Sholes & Glidden typewriter entered commercial use in the 1870s, written work began to separate legibility from the expressive marks of handwriting, and evaluation shifted from visible penmanship toward the quality of thought on the page. AI is doing something similar at a much larger scale. It is stripping away proxies we used to measure human capability and leaving us with the harder task of naming the capability itself.

This book makes 3 claims.

First, AI has made execution cheap, and that exposes how much of school, work, media, and culture had already been organized around compliance and polish instead of judgment, understanding, and real contact.

Second, as execution gets cheaper, value migrates toward judgment, provenance, presence, context, and responsibility, the things that are harder to automate cleanly. Together they form the ability to stand behind something and say, with full accountability, "I chose this."

Third, that migration will not happen on its own. It requires deliberate design in schools, organizations, creative practice, governance, and daily professional life. Cheap execution can create better thinking. It can also create more comfortable drift.

The book moves through 4 parts. Part I holds up the mirror and looks at what AI exposes about performance, trust, and accountability. Part II traces the missing middle: the history, the technology, and the developmental cost of bypassing effort. Part III turns to the human advantage, to context, presence, and good use in practice. Part IV takes up the renegotiation, what institutions owe people when the benefits and burdens of AI fall unevenly.

This book leaves tool instruction aside and turns toward the harder matter of what should never be handed over to a tool.

The mirror is already here. What matters now is what we are willing to see.

INTRODUCTION NOTES

Reuters. "ChatGPT sets record for fastest-growing user base," February 1, 2023, citing a UBS estimate that ChatGPT reached 100 million monthly active users in January 2023. For Clarkesworld's temporary closure, see Time, "Sci-Fi Publishers Are Upset Over Heaps of Unwanted AI-Generated Pitches," February 22, 2023, and Guardian, "Sci-fi publisher Clarkesworld halts pitches amid deluge of AI-generated stories," February 21, 2023. For the typewriter timeline, see Library of Congress, "Production on the Sholes and Glidden Type-Writer Began."

PART I

The Mirror

CHAPTER 1

The Performance of Work

Look at the ordinary documents that fill a working day: the essay, the quarterly business review, the nursing handoff report.

Sometime in early 2024, a Connecticut-based school superintendent, I will call her Jane, told me about a moment that shook her view of education. One of her teachers had received 2 essays about F. Scott Fitzgerald's The Great Gatsby from 2 different students. One was written by a bright, hardworking student over 3 painstaking hours. The other was generated by ChatGPT in about 30 seconds.

The problem was the sameness. Each had a perfect 5-paragraph structure, a clear thesis statement, 3 supporting paragraphs with textual evidence and proper MLA citations. Each concluded by restating the thesis in slightly different words.

If you looked at the metadata, you could see the cheating. But if you looked at the work, the output that teachers had told students was the goal of their learning, you couldn't tell the difference. If a student could spend 3 hours writing an essay indistinguishable from what a machine produces in seconds, what was the assignment teaching?

The answer made her uncomfortable.

The assignment was teaching students to follow a format. To find quotes that supported a predetermined thesis, structure paragraphs according to a rubric, and produce output that looked like analysis without requiring any thinking about the text. The student who used ChatGPT cheated. That's certain. But the task had been cheating students for years, by testing compliance instead of understanding. It rewarded the performance of scholarship over the struggle that produces it.

The classroom problem is the origin story of the modern workplace.

The Corporate Theater

By the time we graduate, the lesson has been learned. Success means figuring out what the evaluator wants and delivering it in the correct format. Then we graduate, and that lesson follows us into the office.

I've sat in boardrooms presenting strategy decks that took weeks to build, slide after slide of data, formatted perfectly, checking every corporate box. I wasn't communicating a vision, as I said I was. I was proving I had done the work, performing "strategy." My effort was aimed at proof, the insight was beside the point, and the proof was accepted without anyone asking whether the insight behind it was real.

I spent years in digital marketing, running campaigns for film studios, non-profits, and consumer brands. The client work that got praised was rarely the most creative or strategically sound. It was the work that came in the most polished format, with the most impressive deck, at the most efficient turnaround. Format was signal; substance was background noise. It took me longer than I'd like to admit to realize how much of what I was producing was performance all the way down.

A project manager at a Fortune 500 company had been writing weekly status reports for 3 years: 3 pages, charts showing project timelines, bullet points summarizing progress, risks flagged in red, action items in yellow. Each one took about 2 hours. In early 2024, she fed her project data into ChatGPT, and 30 seconds later she had

3 pages nearly identical to what she'd been producing.

She edited for accuracy, fixed a few details, and sent it. Total time was 10 minutes. Her director's response was simple. "Thanks for the update. Looks good."

Same as always.

If her director couldn't tell the difference between 2 hours of her work and 10 minutes of AI-assisted work, those hours had been producing something other than work. The honest answer was evidence. She had been producing evidence of effort, and the value was a separate matter that no one was asking about.

The anthropologist David Graeber gave this phenomenon a name in 2018: bullshit jobs. He identified "box tickers," employees whose primary function is to allow an organization to claim credit for work that has no substance behind it. She had become a digital box ticker, assembling evidence that value was being built, one report at a time.

The information transfer itself could have happened in a 5-minute conversation. "Project Alpha is on track. Project Beta is delayed 2 weeks due to vendor issues. Project Gamma needs a decision on scope." But organizations don't reward short conversations. They reward artifacts that demonstrate effort, documents that can be filed, referenced, and forwarded up the chain.

She started paying attention to how much of her work followed this pattern: the weekly team meeting where everyone shared updates they'd already posted in Slack, the monthly steering committee presentations regurgitating data everyone had already seen, the quarterly business reviews synthesizing information from the status reports that had already been synthesized from the data. Layer upon layer of summarization, each requiring human effort, each adding nearly nothing new.

This pattern burns human energy on an industrial scale, crowds out the work that matters, and trains people to perform performance. And when a machine arrives that can produce the appearance without any human energy at all, everyone suddenly has to confront what the appearance was worth.

Health systems are finding a similar dynamic in clinical documentation. Ambient AI documentation tools can generate draft notes or structured summaries from the clinical record and the clinician's narrative. The information still matters, because continuity of care depends on it, and the draft still requires clinical review.

The early evidence points toward reduced documentation burden while leaving clinical judgment intact. Some implementations report savings of several minutes per encounter and reductions in after-hours documentation. What the AI absorbs is the mechanical assembly of known information into a known structure.

That part was never the heart of nursing or medicine. It was often what got in the way of it.

When documentation burden falls, clinicians gain a choice. The recovered time can return to the patient, to the family, to the subtle change in condition that would otherwise be missed, or it can disappear into higher throughput.

The technology does not decide which of those happens. Institutions do.

Law firms tell a version of the same story. Associates spend years learning to draft routine contracts, write standard memos, and format documents according to firm conventions. Much of this follows templates, because the same legal issues arise repeatedly and the same contract terms are negotiated the same way. AI can generate first drafts of routine contracts in minutes, outline standard legal memos, and format documents precisely every time. Firms have been billing for all of it, the formatted work and the strategic work, the performative and the substantive, and the billing rates told the same story they'd always told. But now the story was harder to defend.

An analytics manager I know runs a team producing weekly reports on sales trends, customer behavior, inventory levels, and competitive pricing. The reports go to multiple teams, each wanting them formatted differently, which means her team spends more time reformatting the same underlying analysis than doing the analysis itself. When she finally used AI to automate the reformatting, she discovered something unsettling. The analysis itself took about 45 minutes a week. Everything else was packaging, and the packaging had consumed her team's days.

The same pattern is revealing itself across industries. We have been constructing a world built on the proof of effort, with judgment written off as a luxury we said we couldn't measure. And we built it so thoroughly, so consistently, that most people inside it could no longer tell the difference between the two.

The Blacksmith Test

In the pre-industrial era, a blacksmith judged the fire's heat by its color, felt the resistance of the metal against the hammer, and knew his customer and the purpose the tool would serve. Execution and discernment were combined in every strike.

The factories that arrived in the 19th century severed that connection. They separated thinking from doing, and called the result progress. The cost of the separation took a hundred years to surface, and AI is forcing us to stitch the two back together. There is a deeper history behind that severing, and chapter 4 picks it up. For this chapter, the point is the present moment: we have been operating inside the consequences of an old bargain for so long that we mistook the bargain for the work.

The 2025 Reality Check

By early 2025, the scale of AI's entry into professional work had become undeniable.

Stanford HAI reported that 78 percent of organizations were using AI in some form, up from 55 percent the year before. Anthropic's economic data shows that 77 percent of enterprise API usage involves automation-dominant tasks, and in 8 months, "directive" task delegation in Claude conversations rose from 27 percent to 39 percent. The conversation had moved from "Can AI write?" to "How much work are we delegating, how fast, and who still owns the decision when the system acts?"

The World Economic Forum's 2025 Future of Jobs Report projected that while 92 million jobs might be eliminated by 2030, 170 million new roles would be created, resulting in a net gain of 78 million. These numbers should be held lightly, since economic projections about technological disruption have a poor track record. But the directional point matters here. What's happening is transformation, and replacement is the wrong frame for it.

What gets transformed is more than just the tasks. It's the meaning of the tasks. When an AI can do in seconds what you did in hours, the hours don't disappear. They become available for something else, and we can use that time deliberately or fill it with more of what wasn't working before.

Research from Harvard Business School and BCG on what they call the "Jagged Technological Frontier" found that consultants using AI completed tasks 25 percent faster and produced 40 percent higher quality results on average. But AI acted as a leveler: it raised the floor for lower performers and often homogenized the output. The people who saw their roles grow were the ones who used the saved time to navigate the complex, ambiguous edges of the frontier, the parts the AI got wrong, the decisions the data couldn't make.

That finding contains a quiet revolution. AI changes more than who does the work. It changes which work is worth doing. The activities at the clear center of a knowledge domain, things like summarization, formatting, and the application of known patterns, are being absorbed by the machine, while the activities at the ambiguous edges, where discernment is required and values must be weighed, are becoming the core of what it means to work.

A Necessary Caveat

Output still matters. A clear handoff can protect a patient, a short status update prevents a bad decision, and a well-drafted clause can keep a partnership from collapsing. Trouble starts when output becomes the whole contribution rather than the visible trace of judgment underneath it. AI forces us to ask where the judgment lives.

The Honest Response

When this revelation lands, there are 2 choices.

The first is defensive. Double down, build better detectors, create stricter rules to force humans to keep acting like robots, demand that students prove they wrote their 5-paragraph essays by hand. This path leads to an arms race between detection and evasion that humans will lose.

The second response is more difficult and more honest. If AI can do it, maybe we should let it. And then ask what we should be doing instead.

If the technology can perform, we can stop performing. We can focus on work that requires human discernment, work where the thinking matters more than the format, where presence and struggle add value that pattern-matching cannot replicate.

Jane chose the second path. She put the assignment under pressure. "If AI can complete this assignment in 30 seconds, should we still be assigning it?" For the Gatsby essay, the answer was no.

They realized the old assignment tested whether students could follow instructions, and that understanding the novel had never been directly assessed. So they changed it. Instead of asking for a standard essay on symbolism, they asked students to explain how the green light means something different to 2 characters, then explain why Fitzgerald might have wanted one symbol to carry different meanings. This assignment is harder to automate. AI can generate an answer, but it will be generic, and it will miss the nuance of a student's specific reading. More importantly, the assignment requires the work that matters: reading carefully, forming interpretations, making choices about what to include and what to leave out, standing behind an argument.

The format doesn't matter anymore. The thinking is what gets graded.

A 2024 study by Doshi and Hauser found that while AI helps individual writers produce "better" stories by conventional measures, it reduces the collective diversity of the group. Stories become more polished and more similar, and the homogenization showed up most strongly in the areas where AI is most confident: sentence-level structure, conventional narrative arcs, standard argument forms. These were the areas where human writers diverged most from each other before AI arrived. If we don't change the assignment, we're training an entire generation to think alike.

Jane didn't stop with one assignment. She convened a working group that spent several weeks testing every major assignment in their curriculum against a single question. If a student could complete this using AI, would completing it still accomplish our educational goal?

For many assignments, the answer was no. The assignments were not bad; the educational goal had been implicitly defined as producing the output rather than developing the capability. The working group asked what those assignments were meant to develop, and whether a redesigned version could develop it.

The results were instructive. A history assignment shifted from summarizing an event to arguing, with specific evidence, for why a different decision at a particular juncture would have changed the outcome. In science, students proposed their own experiments, designed the protocol, anticipated confounding variables, and explained what result would falsify the hypothesis. A math assignment sent students to find a problem in their own community that could be modeled, collect actual data, and interpret results. A literature assignment required students to find a passage from the text they disagreed with, explain why, and then argue the strongest possible case for the opposite position.

None of these are easily automated. AI can generate answers, but they will be generic. They won't reflect the specific student's reading, the specific community's data, the specific student's reasoning. The history student who argues a counterfactual has to understand causation, the science student who designs an experiment has to understand methodology, and the math student who finds real data has to understand modeling. The literature student who argues against her own reading learns to hold 2 positions at once, which is one of the most important cognitive skills there is. The assignments require the student to bring something of themselves.

The redesigned curriculum made teaching harder to grade, harder to standardize, and nearly impossible to scale. It also made learning happen. And that was the point.

The Economic Reality

The optimistic version of this argument, the "everyone moves to higher-order thinking" version, glosses over something important.

If the program manager can do her reporting in 10 minutes instead of 2 hours, does her company need as many program managers? The same math haunts the law firm that can draft contracts in seconds, and the analytics team whose actual analysis takes 45 minutes a week.

Strip away the performative layer, and we might find it was the only thing keeping some people employed. And those people are often not the ones at the top of the pay scale. They are the people in the middle: the junior associates, the entry-level analysts, the new graduates who needed the structured work to justify their salaries while they built their judgment.

The MIT economist David Autor has studied this carefully. His research suggests that the automation of routine tasks eliminates more than jobs. It removes the bottom rungs of the ladder. The tasks that used to be entry points into professional development, the work simple enough for a novice but real enough to warrant employment, are precisely the tasks AI absorbs most cleanly.

If the bottom rungs disappear, how do people climb? The junior roles were the crucible where discernment got built, and nobody has proposed a replacement.

The discomfort is real, and the answers are too. Any honest reckoning with AI's impact on work has to sit with both. The performative layer was wasteful, and removing it may also be hard on people in uneven ways.

The nurse who spends less time charting doesn't become less needed; she has more time for the human connection that heals. The teacher who spends less time grading grammar becomes a mentor who guides thinking, while the syntax-correcting goes to the machine.

But this only works if the organizations around those professionals redesign their roles accordingly. If the hospital just hires fewer nurses because the charting is automated, the 25 freed minutes don't benefit anyone. They disappear into the margin.

The transition is a renegotiation of what value means. And renegotiations, even necessary ones, are costly. They require difficult conversations and deliberate choices that institutions rarely make voluntarily until the cost of not making them becomes undeniable.

That renegotiation returns later in the book. But first, a longer look at what the mirror is showing.

The Mirror Stays Up

How much of your work could be automated? Set aside the work you wish you were doing, or the work you think should matter. The work you do: the reports, the meetings, the emails, the presentations.

If AI can do it, what does that tell you? Maybe it tells you the work never mattered and was always performance. Or maybe it points to work that matters, but you've been doing it wrong, following formats when you should have been making choices, chasing efficiency when you should have been chasing insight.

Whichever answer comes, something true is surfacing.

Neil Clarke kept Clarkesworld submissions closed for 3 weeks while he built better filters. Once he reopened, he tried to screen out the bots, but he also started paying closer attention to something else. The human-written submissions sounded like AI could have generated them: stories that followed formulas too neatly, technically competent but saying nothing new.

He had been rejecting those stories for years. Now he could see them for what they were.

Real stories, like good strategy, good teaching, and honest care, require something templates can't provide. A particular way of seeing. A specific voice. Choices that reflect discernment and risk. The submissions that worked weren't the ones that followed the rules better. They were the ones that earned the right to break them.

Goodhart's Law and the Proxy Trap

There is a principle in economics called Goodhart's Law: when a measure becomes a target, it ceases to be a good measure. First articulated by the British economist Charles Goodhart in the context of monetary policy, it applies almost universally to any system where performance is evaluated according to measurable proxies.

The 5-paragraph essay is Goodhart's Law applied to writing education. The measures (clear thesis, 3 supporting points, restated conclusion) were chosen because they were gradeable. They never captured what good writing is. Once they became targets, students aimed for them, and the measure stopped capturing what was supposed to be measured.

The status report follows the same pattern. The measure of being an engaged professional had become whether you produced a regular written update, a proxy chosen because it was observable, never because it captured what good communication is. Every corporate KPI you've ever been evaluated on follows this logic. Someone decided the valued outcome was too hard to measure directly, chose a proxy, and then the proxy became the target.

AI forces a confrontation with this pattern. When AI can hit almost any metric as well as or better than humans, the metrics that were proxies reveal themselves as such, and the test they always should have prompted becomes unavoidable. What is this metric actually for? And beneath that, what were we trying to accomplish before we confused the measure with the thing?

What the Mirror Asks

AI is revealing something difficult. The processes we called thinking, creating, and working had always involved the offloading of cognitive labor to external tools, and the difficulty, as a result, of saying precisely what the distinctively human contribution is. The chapters that follow trace that argument across the workplace, the historical record, and the technologies themselves. Chapter 4 picks up the institutional history that produced the bargain we are now renegotiating.

The revelation is uncomfortable. It is also generative, and it is playing out in one domain after another.

Different domains will arrive at different answers. Surgeons will locate it in the clinical relationship, the tactile judgment, the integration of knowledge in the moment of intervention. Teachers will find it in reading what a particular student needs, the thing no data system captures. For leaders it is judgment under uncertainty, the moral weight of consequential decisions, and the trust-building that makes organizations function as more than coordinated self-interest; for writers, it is the truth they are trying to tell, what they see and want the reader to see, and the accountability to that vision that makes the work worth making.

Each domain will find its own answer, but the answers share a common structure. The irreducibly human contributions are the ones that require a person to be fully present, accountable, and committed to something beyond the performance of the role. AI reveals this by handling everything else well enough to make the distinction visible.

Clarity is uncomfortable, and it is also essential. The discomfort of this clarity is worth sitting with, because the alternative is to fill the space AI opens with more production and more processed output, without ever asking what the processing is for.

AI is an invitation, and treating it as a verdict misreads it. It is the invitation to ask, finally and honestly, what human beings are doing here.

The work that comes next is ours to choose.

The Monday Morning Test

Before you close this chapter, try something. Look at your calendar for next week and pick 3 tasks: a report, a meeting, a deliverable. For each one, ask yourself whether AI could do this in 30 seconds, and whether the result would be meaningfully different from what you produce. If the answer is no, the task is packaging. It may still need doing, but stop confusing it with your actual contribution. If the answer is yes, if something about your judgment, your knowledge of the people involved, your reading of the situation makes your version different from what a machine would generate, that is the work. Protect it. The distinction between packaging and contribution is the most important line you can draw in your professional life right now, and few of the professionals I work with have ever drawn it.

CHAPTER 1 NOTES

The superintendent, called Jane here, is an anonymized portrait; identifying details have been changed, consistent with the Note on Sources and Method at the front of this book.

Graeber, David. Bullshit Jobs: A Theory. Simon & Schuster, 2018. Graeber's taxonomy of "bullshit jobs" includes five types: flunkies, goons, duct tapers, box tickers, and taskmasters. Box tickers are specifically defined as employees who allow organizations to claim credit for work with no substance behind it.

For documentation of AI-assisted clinical documentation, see Microsoft, "DAX Copilot: New customization options and AI capabilities for even greater productivity," August 8, 2024, reporting an average of 5 minutes saved per clinician encounter; Haberle, Tyler, et al. "The impact of Nuance DAX ambient listening AI documentation: a cohort study." Journal of the American Medical Informatics Association 31, no. 4 (2024): 975-979. doi:10.1093/jamia/ocae022; and Wendt, Staci J., Catherine T. Dinh, Michael G. Sutcliffe, Kyle Jones, James M. Scanlan, and J. Scott Smitherman. "Deploying Ambient Clinical Intelligence to Improve Care: A Research Article Assessing the Impact of Nuance DAX on Documentation Burden and Burnout." Future Healthcare Journal, 2025. doi:10.1016/j.fhj.2025.100450.

Stanford University Human-Centered Artificial Intelligence. AI Index Report 2025. Stanford HAI, 2025. The AI Index documents both adoption rates and capability benchmarks across industries and geographies.

Anthropic. "Anthropic Economic Index report: Uneven geographic and enterprise adoption of Claude," September 15, 2025. The report documents directive conversations rising from 27% to 39% and notes automation-heavy API usage.

World Economic Forum. Future of Jobs Report 2025. WEF, 2025. The report's job displacement and creation projections represent modeled scenarios, not certainties, and are subject to wide confidence intervals.

Dell'Acqua, Fabrizio, et al. "Navigating the Jagged Technological Frontier: Field Experimental Evidence of the Effects of AI on Knowledge Worker Productivity and Quality." Harvard Business School Working Paper 24-013, 2023.

Doshi, Anil R., and Oliver P. Hauser. "Generative AI Enhances Individual Creativity But Reduces the Collective Diversity of Novel Content." Science Advances 10, no. 28 (2024): eadn5290.

Autor, David H. "Applying AI to Rebuild Middle Class Jobs." MIT Work of the Future, 2024. Autor's concern about "hollow ladders" refers specifically to the removal of entry-level cognitive work that served as professional development infrastructure.

CHAPTER 2

The Trust Collapse

A man named Jason Allen walked into the Colorado State Fair in late summer 2022 and started a fire that is still burning. He hadn't brought matches or a canvas. He brought a piece of digital art titled Theatre D'opera Spatial: 3 figures in elaborate robes standing in a circular baroque chamber, a massive circular opening bathing the scene in golden light, something like a space opera painted by a Renaissance master.

The judges awarded Allen first place in the Digital Arts category. Then the world found out how he made it.

Allen had used Midjourney, an AI image generator. He had spent weeks refining text prompts, generating hundreds of variations, selecting and upscaling the best outputs, and he was transparent about it. He had titled the submission "via Midjourney." But when the news spread, the backlash was immediate and fierce. Artists called it cheating. Critics called it the death of art.

The controversy was about authenticity, about what lets an audience trust that a work is real and that someone stands behind it.

We are now in the middle of a trust collapse, and it extends far beyond art.

The Collapse of the Hand-Made Signal

For centuries, people operated under a simple unspoken agreement: authenticity equals human labor. A painting was revered because an artist held the brush. A novel was worth reading because a writer had bled over the keyboard. We valued the sweat equity, assuming that if something was beautiful or useful, it must have required human suffering and skill.

Jason Allen broke that agreement. He produced something beautiful without the manual toil we associate with beauty, and because he bypassed the process, many felt he hadn't earned the result.

Technology had chipped away at execution-as-value before. When photography arrived, painters panicked, arguing that capturing an image on a chemical plate wasn't art because the machine did the work. Eventually we accepted photography and located the art where it had moved: the choice of framing, lighting, and timing. The artist went from being a maker of marks to a selector of moments.

AI creates a similar shift, but more extreme. It democratizes execution to near-zero cost. When anyone can generate a baroque space opera or write a sonnet or build a website, execution stops being a reliable signal of authenticity. You can no longer look at a piece of work and assume a human spent 80 hours on it.

This loss creates a vacuum of trust. Call it proxy failure. We measured effort and difficulty as a stand-in for sincerity, and now the effort is free.

You feel the proxy failure in small moments before you notice it in large ones. A friend sends you a text after a disagreement, and it's unusually well-crafted. It takes responsibility without groveling, validates your feelings without being sycophantic, avoids defensiveness while maintaining dignity. The punctuation is clean, the emotional tone warm but measured. It's perfect, and the perfection is what unsettles you. You find yourself wondering: did my friend write this, or did they paste our argument into ChatGPT and send me the output? The words are right, but you can't tell whether a person is behind them. The proxy has failed. The sweat was the signal, and the signal is gone.

In 2023, the online book market experienced a surge of AI-generated titles. People were using automation scripts to generate entire books, create AI-generated covers, and upload them to publishing platforms by the hundreds: romance novels, children's stories, self-help books. The authors listed on the covers didn't exist, the advice was generic, the stories often incoherent. No human consciousness said, "I believe this story is worth telling," only a desire to extract revenue from an algorithm.

These books didn't have an AI problem. They had a nobody's home problem. The text was machine-generated, the recommendation algorithmic, the purchase human, and the human felt cheated. If we pass off machine execution as human effort, trust evaporates. We create a ghost product that may look like a book, but there's no one standing behind it.

The Beatles Answer

Compare that to The Beatles. In late 2023, Paul McCartney and Ringo Starr released Now and Then, billed as the Beatles' final song. The track was built around a demo John Lennon had recorded on a cassette tape in the late 1970s, unusable for decades because the piano was too loud, the tape hiss overwhelming, Lennon's voice buried. Peter Jackson's team had developed advanced AI audio separation technology for the Get Back documentary, and they used it to surgically isolate Lennon's vocal from the noise.

With Lennon's voice rescued, Paul and Ringo finished the song, adding bass, drums, strings, and backing vocals, collaborating with their deceased friend across time, mediated by a machine. The song won a Grammy. People wept.

We did not attack Paul McCartney the way we attacked Jason Allen, because he told us the truth. McCartney didn't hide the AI. He gave interviews explaining exactly how it worked, making clear that the technology didn't write the song or sing the notes but acted as a tool to reveal something human that was already there. The authenticity of Now and Then didn't come from hand-made purity. It came from transparency. People trusted the song because they trusted the humans who made the choices. Since almost everything has now been touched by a machine somewhere in its production, what people need to know when they encounter AI-assisted work is whether a human stands behind it. Is there a consciousness that chose this, took responsibility for it, and will answer for it?

When the answer is yes, trust survives. When the answer is no, or unclear, trust collapses.

Collaborative Authenticity

Rie Kudan won Japan's most prestigious literary award, the Akutagawa Prize, in 2024, for a novel exploring a near-future world where AI had become deeply integrated into daily life. After winning, she revealed that about 5 percent of the novel was written by ChatGPT. The press gasped. The judges didn't care, and one committee member went on record saying that what really mattered was whether the work had something meaningful to express.

Kudan had used AI thematically. She employed the machine's tendency toward vague, pleasant language as a texture in the narrative, representing the clouded thinking of characters confronting justice, and she spoke to the AI about thoughts she couldn't tell anyone else, then incorporated the responses to shape a protagonist's isolation. The machine's voice was being curated to serve a specific human artistic vision. She stood behind every word, even the machine-generated ones, because she had chosen them for a purpose.

The machine didn't write the book. Rie Kudan wrote the book using the machine.

The Voice That Wasn't There

In April 2023, an anonymous user posting as Ghostwriter977 released a song featuring AI-cloned vocals of Drake and The Weeknd, first to streaming platforms and then virally on TikTok. The track spread across TikTok, Spotify, and YouTube before Universal Music Group had it removed. The song sounded right, the voices were convincing, and listeners who didn't know it was AI-generated responded to it as music. They felt things.

Then came the disclosure, and with it the most telling institutional response of the entire AI-art controversy. Harvey Mason Jr., CEO of the Recording Academy, initially told reporters the track was "absolutely eligible" for a Grammy "because it was written by a human." Days later, he reversed himself. Ghostwriter responded by re-uploading the track with his own voice replacing the AI clones, raising an issue that exposed the whole system. If the melody, lyrics, and production are identical, is it the same song? Something made the first version ineligible that the second version fixed.

The answer is that the first version broke a proxy no one had examined. We assume that a voice belongs to the person it sounds like. The human voice is something close to a signature, an indexical sign, like smoke pointing to fire, and AI voice cloning broke that index. A UC Berkeley study in 2025 found that people are "poorly equipped to identify AI-generated voice clones." The proxy we relied on, if it sounds like Drake, Drake is behind it, no longer holds.

What made this case devastating was that the audience's experience was real. People enjoyed the song. The crisis began with the disclosure: listeners had already judged the music and liked it, and what broke was the attribution.

The Broken Circuit

In 2024, researchers Brian Porter and Edouard Machery ran an experiment at the University of Pittsburgh. They gathered readers and showed them poems, some by Shakespeare, Dickinson, Whitman, and Byron, others generated by ChatGPT in the style of those same poets, without telling participants which was which. Then they asked for ratings on quality and emotional resonance.

The results were humiliating.

AI-generated poems scored higher. Readers rated the machine-written verses as more rhythmic, more beautiful, more inspiring than the actual poets, and when asked to identify the human authors, participants performed worse than random chance.

Then the researchers ran a second experiment. They told a new group which poems came from humans and which from AI.

Once readers knew a poem came from a machine, they rated it lower across 13 of 14 quality measures. The same rhythm they had praised now felt mechanical, the moving emotion became shallow, the allure disappeared. Nothing about the words had changed. Only the circuit.

The readers had learned there was no consciousness on the other end of the line. The mind they thought they were communing with had never known love or grief or the smell of rain, and what they were reading was a statistical prediction of how a poem about those things ought to sound. The connection snapped, and the value went with it.

Meaning lives in the connection between two minds, which is why identical words can carry it one day and lose it the next. Once we remove the human from the creation, we change more than the work's origin. We break the communicative circuit that makes the work matter.

My PhD thesis research found the same pattern across literary and visual art. The perceived value of a work changes dramatically based on what audiences believe about its origin. The work itself may be identical; the attribution determines the aesthetic experience.

Hans Robert Jauss, the reception theorist, called this the "Horizon of Expectation." We approach every creative work with an implicit assumption: a human mind that experienced something is trying to communicate that experience to me. Once that assumption is violated, the horizon collapses. The shock is about the fundamental communicative contract between creator and audience.

Across the cases I examined, authenticity was migrating from a property of the object (made by human hands, using traditional methods) to a property of the relationship between creator and work: this person understood what they were making and is accountable for it. What audiences require is the felt sense that a consciousness was present in the making, that choices were made by someone with something at stake.

The Provenance System

This shift from "made by hand" to "vouched for by a person" is becoming a technical and legal reality.

The Coalition for Content Provenance and Authenticity, known as C2PA, has developed a technical standard called Content Credentials, and Adobe, Microsoft, Leica, Nikon, and other major players are pushing it as an industry baseline. The standard embeds cryptographically signed metadata into images, videos, and documents, recording the chain of custody from creation through any modifications.

The EU AI Act now treats output transparency for AI-generated or manipulated content as a formal legal obligation, including deepfakes and text intended to influence public opinion.

Yet practical implementation remains shaky. A Washington Post investigation found that major social platforms often failed to preserve or display Content Credentials metadata when users uploaded files. The signature got stripped in transit. The technical solution exists; the platform incentives to implement it consistently do not yet align.

This gap between technical capability and consistent implementation is where much of the trust problem lives. The tools to establish provenance are becoming real. The will to use them lags behind.

What Trust Now Requires

The old authenticity signal was effort. The new one is provenance. The reader, patient, client, student, or audience member needs to know that a human being understood the work, made choices inside it, and can answer for the result. Trust now depends less on whether a human performed every step and more on whether a human stands behind the work that leaves their hands.

Building Your Provenance Habit

Three practices build trust when the sweat signal is broken.

Disclose before you are asked. When you use AI to help produce something, say so, without apology and factually. "I used AI to draft the initial research summary, then spent 2 hours restructuring the argument and verifying the sources." That sentence builds more trust than silence ever could.

Show your choices. The value of your work now lives in what you selected, what you cut, what you changed. Make those decisions visible.

Sign only what you have inhabited. Read every word that goes out under your name, and check the facts you did not generate. If you would not defend it in a room full of skeptics, do not send it.

Provenance is a daily practice of being the person who stands behind the work.

CHAPTER 2 NOTES

For documentation of AI-generated book flooding on retail platforms, see Cole, Samantha. "Amazon Is Selling Books Generated by AI That Falsely List Authors' Names." 404 Media / Vice, August 2023; and Reuters reporting from February 2023 onward documenting the rise of AI-generated titles on Kindle Unlimited and the Amazon storefront.

Tsioulcas, Anastasia. "The Beatles release a new song, with a little help from AI." NPR, November 4, 2023. For analysis of audience reception, see subsequent criticism and commentary in Rolling Stone, Pitchfork, and academic reception studies.

For analysis of Rie Kudan's Akutagawa Prize and the AI disclosure controversy, see reporting in The Japan Times, Nippon.com, and subsequent academic treatment of the case as a milestone in AI-assisted literary fiction.

Porter, Brian, and Edouard Machery. "AI-Generated Poetry Is Indistinguishable from Human-Written Poetry and Is Rated More Favorably." Scientific Reports 14 (2024): 26133. For the Ghostwriter/Drake/Weeknd case, see Snapes, Laura, The Guardian, April 18, 2023, and Aswad, Jem, Variety, September 8, 2023. For voice-clone detection, see Barrington, Sarah, Emily A. Cooper, and Hany Farid. "People are poorly equipped to detect AI-powered voice clones." Scientific Reports 15 (2025): 11004.

Jauss, Hans Robert. Toward an Aesthetic of Reception. Translated by Timothy Bahti. University of Minnesota Press, 1982. The concept of the "Horizon of Expectation" describes the pre-programmed interpretive framework audiences bring to texts based on genre conventions and cultural norms.

Coalition for Content Provenance and Authenticity. C2PA Technical Specification, Version 2.4, April 2026. Available at spec.c2pa.org.

European Union. Regulation (EU) 2024/1689, Artificial Intelligence Act, Article 50, 2024. See also European Commission, "Draft guidelines on the implementation of the transparency obligations for certain AI systems under Article 50 of the AI Act," May 8, 2026.

Washington Post. "We uploaded a fake video to 8 social apps. Only one told users it wasn't real." October 22, 2025. The investigation tested how major platforms handled AI-video disclosure and Content Credentials metadata.

McBain, Ryan K., et al. "Use of Generative AI for Mental Health Advice Among US Adolescents and Young Adults." JAMA Network Open 8, no. 11 (2025): e2542281. doi:10.1001/jamanetworkopen.2025.42281.

CHAPTER 3

The Signature

A lawyer named Steven Schwartz sat in a federal courtroom in Manhattan in June 2023 and watched his career disintegrate. He had represented a man suing an airline for injuries sustained from a serving cart, a standard personal injury case, and he had filed a brief citing more than half a dozen previous court decisions, complete with case names, quotes from judges, and case numbers.

There was one problem. The cases didn't exist.

When pressed, Schwartz admitted the truth. He had used ChatGPT to do his legal research, and the chatbot had hallucinated the cases out of thin air. It even wrote convincing, legally sound paragraphs that sounded like federal judges writing about airline liability. Schwartz had asked ChatGPT directly whether the cases were real. The machine replied, "Yes."

He was a tired professional trying to use a new tool to be more efficient. Malice was never the issue. He made a fundamental error about what the tool was: he thought he was using a search engine when he was using a sentence completion system.

He made a professional mistake, and the technical mistake was downstream of it. He put his name on a document he hadn't verified, signed a brief containing assertions he didn't understand, and abdicated the one duty that defined his job, which was to stand behind his work.

The judge sanctioned him and the press mocked him, but the rest of us shouldn't laugh too hard. Steven Schwartz is the ghost of a future many professionals are drifting toward, and his temptation is now everyone's: the seduction of handing over the burden of knowing to something that can simulate knowing very convincingly.

The Decoupling Problem

Our current trajectory is producing a widening gap between our power to generate work and our power to understand it. You can generate a marketing strategy without understanding your customer, a lesson plan without understanding pedagogy, a legal brief without knowing the law.

A crisis of accountability follows. When the machine does the work, someone still has to be responsible for the mistake, the bias, and the harm. The AI system does not answer for them. The humans do: the one who used it, the one who approved the output, the one who signed the document.

A professional is no longer the person who does the work. A professional is the person who decides the work is good enough to be signed.

"Knowing" a domain has historically entailed more than recall. A physician who knows medicine has cultivated, through years of clinical exposure, a kind of perception. They notice things: a patient's hesitation before answering something sensitive, the slight incongruence between the stated complaint and the body language. This perceptual layer, built through immersion and repeated engagement with reality, is what allows the physician to catch the error in the AI's recommendation, and it lives in the practitioner rather than in any reference book. Outsource the cognitive work that builds that perceptual layer, and we prevent its development. The McKinsey analysis of skills likely to remain "uniquely human" through 2030 identified specifically those rooted in social and emotional intelligence, interpersonal conflict resolution and design thinking, which depend on empathy, creativity, and contextual understanding. These skills come from engagement rather than information, from showing up in difficult situations repeatedly until your nervous system learns to handle them.

The Authority Illusion

If you delegate thinking to an AI, you are delegating to an entity that does not care whether it is right, whether you get sued, or whether a patient dies. The system generates the next most plausible token in a sequence. That's what it's doing when it sounds most authoritative.

I call this the Authority Illusion. We assume that because the output looks professional, the process behind it was rigorous: perfect grammar means the facts are true, clean code means the system is secure. We transfer the burden of thinking to a machine that has none.

An engineer using AI to calculate load-bearing requirements faces this directly. If the algorithm hallucinates a material property and the bridge cracks, the engineer who stamped the blueprints is accountable. The algorithm doesn't go to prison.

You can outsource the labor. You cannot outsource the liability.

Professional licensing has always existed on exactly this basis. The physician is licensed because they have demonstrated the ability to exercise discretion responsibly in high-stakes situations, and the license is a public commitment: this person can be held accountable. When AI enters the picture, it does not change the underlying principle. Institutions governing professional practice must update their standards to reflect the new conditions.

The Signature Test: 5 Ownership Checks

Over years of working with educators, executives, and creative professionals, I've developed a set of checks that work as a lens for accountability. They are designed to force a pause before work leaves under your name, before you hand your signature to something you haven't fully inhabited. If any answer is no, the work is not ready to be signed.

What am I trying to accomplish, and why? Most of the professionals I watch skip this one. We rush to the tool before we've defined the problem, asking AI to "write a blog post" without deciding what we want to say. Skip purpose, and the tool drives the car. Being clear about purpose lets you define where you end and the machine begins.

What requires my judgment and what doesn't? Formatting a bibliography doesn't require human judgment, and organizing a spreadsheet doesn't need empathy. But writing a difficult email to an angry client involves weighing legal risk against relationship risk, reading the subtext of someone's complaint, deciding how much humility to show. We get into trouble when we treat discretionary tasks as mechanical ones. If the decision involves ambiguity, values, or human relationships, your hands stay on the wheel.

Do I understand what the AI produced? The Steven Schwartz Rule says if you cannot verify the output, you cannot use the output. A non-coder using AI to write Python, a non-writer getting AI to draft an article: useful in appearance, dangerous in practice. You need to learn enough to verify the work, or find a human expert who can. There is no third option that doesn't eventually fail.

Can I explain and defend every choice, in public if I have to? If someone points to a specific paragraph in your report and asks why you recommended this strategy, and your internal answer is "I don't know, that's what the AI came up with," you've failed. If your answer is "I started with AI-generated options, but I chose this one because it aligns with our Q3 goals and minimizes supply chain risk," you're the author. The judgment is yours.

Would I be proud to show my process? If you feel a twinge of shame about how you produced the work, listen to that feeling. It's data. Use AI to fake expertise you don't have, and you feel the urge to hide the process and to blur what the machine contributed. When you use AI as a legitimate force multiplier, a research assistant, a brainstorming partner, a rough drafter, you can be transparent about it, and the human contribution stays visible. Transparency is the antidote to impostor syndrome.

These checks are more of a practice than a compliance exercise. Applied consistently, they build professional ownership. The discipline is asking, before every piece of work leaves your hands, whether you are really prepared to stand behind it.

The Human in the Lead

The technology industry has produced a popular concept called "Human in the Loop." The idea is that AI can be used safely as long as a human reviews the output before it goes live. In practice, it often falls short.

Humans are poor monitors of machines that almost always work. When a system is reliable 90 percent of the time, we stop checking. We scan the document, our eyes glaze, and we hit approve without much thought. Human in the Loop collapses into Human Asleep at the Switch.

The better frame is Human in the Lead. That means moving beyond reviewing output to directing the process: setting the strategy, defining success criteria, challenging the machine's assumptions, actively looking for what is wrong, treating AI like a bright but hallucination-prone junior colleague.

The Wharton Human-AI Research Center published findings in 2025 describing what high-performing AI integrators do. The high performers spent more time editing, interrogating, and redirecting than generating. They treated AI outputs as starting points for their own thinking, and they engaged with the tool adversarially, looking for what was wrong or incomplete before deciding what to keep.

This posture, productive skepticism toward AI outputs,

verification rather than rejection, is the core discipline this moment demands. It takes more effort than rubber-stamping an AI output, but it produces work that is really owned, work that can be defended, work that builds the practitioner's discernment instead of eroding it.

Professional Standards and Governance

Accountability is individual and institutional.

Professional licensing exists because some activities carry consequences severe enough to require demonstrated competence and accountability. Medicine, law, engineering, and architecture all developed audit trails, liability frameworks, peer review, and standards of practice over decades, because the cost of failure is borne by people who cannot always protect themselves.

AI is arriving faster than those institutions can adapt. In late 2023, Sports Illustrated was caught publishing product reviews written by AI authors with AI-generated headshots and fake biographies. No actual human was verifying the products, and when the deception became public, the reputation of a century-old magazine evaporated in days.

Organizations will use AI. What matters is whether they build governance structures that match the accountability the situation requires. Who reviews critical AI outputs? Who has authority to override? Who is named when something goes wrong? What's the audit trail?

These are Monday morning checks. Organizations that answer them honestly, before they need to, will be the ones that survive.

The Signature as a Promise

Our inboxes, feeds, and attention are flooding with synthetic content, and in that flood, the only thing that will carry lasting value is the human signature.

I mean the metaphysical signature, the personal guarantee that a human being has engaged with this reality and will vouch for it. When you sign a document, you're making a promise: I saw this, I thought about this, I believe this. That promise is the foundation of civilization. It's why we can sign contracts, trust a diagnosis, follow a leader. If we allow AI to dilute that promise, getting into the habit of signing things we didn't read and validating things we don't understand, we risk the collapse of the trust networks that hold society together.

Steven Schwartz lost his standing because he broke his promise to the court. The chatbot was the instrument. The broken promise was his.

We must become more protective of our signatures, more intentional about what we're willing to vouch for, more willing to slow down, verify, and wrestle with the material. This will make us slower than the machines, less efficient than those who automate everything.

It will make us trustworthy. And in an age where trust is the scarcest resource there is, trustworthiness is the one advantage that compounds.

CHAPTER 3 NOTES

Weiser, Benjamin. "Here's What Happens When Your Lawyer Uses ChatGPT." The New York Times, May 27, 2023. The case was Mata v. Avianca, heard before Judge P. Kevin Castel in the Southern District of New York.

McKinsey Global Institute. "Agents, Robots, and Us: Skill Partnerships in the Age of AI." McKinsey & Company, November 2025. The analysis drew on 11 million job postings and the European Skills, Competences, Qualifications and Occupations taxonomy.

Korst, J., et al. "Accountable Acceleration: Gen AI Fast-Tracks Into the Enterprise." Wharton Human-AI Research and GBK Collective, 2025. The study examined differential outcomes between high-performing and lower-performing AI integrators in enterprise knowledge work settings.

For documentation of the Sports Illustrated AI content controversy, see Harrison Dupré, Maggie. "Sports Illustrated Published Articles by Fake, AI-Generated

Writers." Futurism, November 27, 2023.

PART II

The Missing Middle

CHAPTER 4

The Machine Human

We did not arrive at the performance crisis described in chapter 1 by accident. Over more than a century, schools and workplaces learned to solve real problems through standardization. Literacy spread, production scaled, and institutions became easier to manage. The same bargain also trained people to treat reliability as the center of human value. AI is making that bargain visible.

In 1911, a mechanical engineer named Frederick Winslow Taylor published a monograph that would quietly remodel the human spirit: The Principles of Scientific Management. Taylor had spent decades in steel mills and factories with a stopwatch, observing workers as they shoveled coal and carried pig iron. He wasn't interested in their craftsmanship, their discretion, or their experience.

Taylor believed that for every task, there was one best way. He calculated the exact angle a shovel should enter a pile of coal and the optimal number of steps between pile and furnace, stripping away the worker's autonomy and replacing it with a procedure.

Taylor wrote a sentence that became the operating system for the 20th century. "In the past the man has been first; in the future the system must be first."

It worked. Productivity exploded. The assembly line, championed by Henry Ford 2 years later, cut the time to build a car from 12 hours to 90 minutes. The standard of living rose, and on the back of this philosophy America built skyscrapers, interstate highways, and a middle class.

There was also a price, a bargain that treated reliability as human value. Then we applied the logic of the factory to everything else.

The Architecture of Compliance

The factory model didn't begin with Taylor. To sustain a world of scientific management, someone needed a steady supply of humans trained to accept standardized inputs, and that supply chain ran through the schoolhouse.

In 1843, Horace Mann, the Secretary of the Massachusetts Board of Education, traveled to Prussia looking for a way to organize the chaotic, unequal terrain of American schooling. American schools at the time were a patchwork: one-room schoolhouses in rural districts, urban academies for the wealthy, denominational schools serving immigrant communities, and almost no shared standard for what a child should know at any given age. Mann wanted a coherent system, universal in reach and publicly funded.

He found it in Prussia, a centralized system built after the kingdom's early 19th-century military and administrative crisis. Mann admired its public funding, teacher training, common curriculum, and order. The same structure also fit the needs of a state that valued discipline, standardized instruction, and civic formation across a whole population. Whatever Mann's democratic intentions, the model carried a logic of compliance that later industrial institutions could use easily.

Mann admired the order; he may not have fully reckoned with what produced it. He returned to Massachusetts and began advocating for an American version, and by 1852, Massachusetts had passed the country's first compulsory school attendance law. The other states followed within decades.

The Prussian model was a masterpiece of efficiency. It introduced age-based grading, grouping children by production date instead of mastery. The ringing bell trained the nervous system to move on command, and the rows of desks pointed every eye toward a central authority: a factory for attention, before there were many factories to feed.

The hidden curriculum that traveled along with this structure was compliance. Children learned that learning meant receiving information from a teacher who held the right answers, and that there was one right answer, printed in the back of the teacher's book. Collaboration was cheating. Curiosity that wandered off the day's lesson was a discipline problem. Your value was determined by how well you could reproduce the standard output, and by the time a child completed 12 years of this, the habits were inscribed deep enough to shape adult life.

Mann's intentions were largely noble. He wanted universal literacy and a literate citizenry, and he believed, with reason, that a public school system was a precondition for democracy. The structure he imported, however, came with its own logic, and once installed, that logic served whoever needed it. The growing American industrial economy needed workers who would arrive on time, accept supervision, and follow procedure. The Prussian model fit that need.

The cost of this design is measurable. George Land, a researcher who had developed creative thinking assessments for NASA engineers, applied the same instruments to children. He tested 1,600 children at age 5, then again at 10, and again at 15. At 5, 98 percent of them scored at the "genius level" for divergent thinking, the ability to imagine multiple possibilities, hold contradictions in mind, and generate novel solutions. By 10, that number had dropped to 30 percent. By 15, 12 percent. Among adults, only 2 percent qualified. Land read the pattern as evidence that schooling and other social pressures suppress divergent thinking over time, though the study can't carry a full causal explanation on its own. Family, culture, testing regimes, and adolescence itself shape the decline too, and schooling is one contributor among several. That's caveat enough, and warning enough.

Standardization delivered real gains. Millions of Americans learned to read who might not have otherwise, and scale made basic education accessible across vast distances. The cost was real, and the gains were real too. Mann built the system, and we never updated it.

By the time Taylor arrived, the workforce had already been preconditioned for his message. Students who had spent 12 years sitting in rows, moving at the bell, and following instructions were ready to stand on assembly lines, move at the whistle, and follow uniform operating procedures. School produced the machine parts. The factory consumed them.

We called humans "resources," and the word has haunted us ever since. Resources are inputs. You assign them, consume them, calculate their cost per unit, squeeze them for output, and replace them when they wear out.

What the Industrial Bargain Cost

For the better part of a century, this bargain held, the way an old lease holds: nobody rereads the terms until something breaks. You accept standardization, perform compliance, and suppress the parts of yourself that don't fit the system, and in return you get stability, a paycheck, and a defined role.

The psychological cost was immense, and mostly invisible. Frederick Herzberg's research in the 1950s and 1960s set out to discover what motivated people at work. The conventional wisdom of the era was that motivation came from compensation, working conditions, and supervisory style: pay people enough, treat them fairly, and they will produce. Herzberg's findings overturned that assumption.

In the original empirical study, Herzberg and his colleagues interviewed 203 engineers and accountants in the Pittsburgh area about incidents that made them feel especially good or bad at work. The pattern that emerged surprised even him. The factors that produced satisfaction were entirely different from the factors that produced dissatisfaction. They turned out to be 2 different scales running in parallel.

Herzberg called the first set hygiene factors: pay, benefits, working conditions, company policies, supervisory style, job security. Get these wrong, and people are miserable. Get them right, and misery falls, but motivation still has to come from somewhere else. Hygiene factors prevent dissatisfaction. They do not produce engagement.

The factors that produced engagement, Herzberg called motivators. Achievement. Recognition for doing real work. The work itself, when it was meaningful and required judgment. Responsibility for outcomes, and the opportunity to grow. None of these had to do with the conditions of employment. All of them had to do with what the person was doing, and whether the doing of it allowed them to feel like a person.

The Taylorist workplace methodically eliminated the motivators. The whole point of scientific management was to remove judgment from the worker, define the task narrowly enough that it could be performed identically by any interchangeable human, and replace intrinsic engagement with external reward. What was left was hygiene. People in heavily standardized jobs were, by Herzberg's account, contained, manageable, compliant, and never within reach of the kind of motivation that produces genuine commitment to a thing worth doing.

The chronic misalignment between what humans need to thrive and how Taylorist organizations measured them produced burnout long before AI arrived. The 2023 Gallup Global Workplace Report found that only 23 percent of employees worldwide described themselves as engaged at work, which leaves the other 77 percent going through the motions, doing just enough to maintain their position, withholding the very capacities, judgment, creativity, and care, that the work increasingly required.

The bargain was reliability for stability, suppressed humanity for a paycheck, and it held while the work could be reduced to procedure. By the late 20th century, the work could no longer be reduced to procedure, but the operating logic of organizations remained the same. That mismatch is the inheritance we now have to confront.

The Judgment Gap

AI exposes the flaw in the Taylorist bargain with a finality that nothing else could achieve. If the goal is machine-like performance, the machine wins every time. AI never gets tired, never asks for a raise, never has a bad day. It can process information faster, remember everything, and apply rules with a consistency that Taylor could only dream of.

This collision, an education and corporate culture designed to produce standardized humans crashing into a technology that makes standardized humans obsolete, creates what I call the Judgment Gap.

On one side, AI forms a rising floor of "good enough" execution. On the other sits the ceiling of human expertise, which requires deep judgment, context, and responsibility. The gap between them is where the danger lies, because that gap is where junior work used to happen: the entry-level coding, the routine copywriting, the document review, the training ground where humans struggled, failed, learned, and eventually developed discernment.

If AI absorbs this middle ground, the pipeline that built experts closes.

MIT economist David Autor puts it precisely. AI "commodifies expertise." It allows someone with moderate knowledge to perform at a higher level by using the tool. A nurse practitioner equipped with sophisticated diagnostic AI might make assessments that previously required a specialist, and a junior developer with an AI system might write code that used to require a senior architect.

But to use the tool well, you need enough sense to know when it is wrong. If the nurse practitioner blindly trusts the AI, she's a liability rather than an expert, and if the junior developer pastes code he doesn't understand, he's a security risk rather than an architect. The value shifts entirely toward meta-cognition: thinking about thinking, auditing the machine, interrogating its output.

The Knowledge Pipeline

Professional expertise in most fields is built through an apprenticeship pipeline, even when it doesn't look like one. Law associates do junior work under senior supervision before becoming partners, medical residents rotate through services under attending physicians before practicing independently, and junior engineers work on components before leading systems. The pipeline moves people from defined tasks toward ambiguous discretion over years.

The pipeline works because the junior work is developmental. It builds domain knowledge through repetition and variation, discernment through exposure to the gap between theory and practice, and professional intuition through the slow accumulation of specific, concrete cases. Strip away the junior work and you strip away the pipeline.

The Microsoft New Future of Work Report 2025 documented this concern explicitly. When AI automates entry-level cognitive work, organizations risk losing the "tacit skills that drive long-term innovation" because workers no longer develop them through hands-on learning. The economy gains efficiency in the short run. It potentially loses capability in the long run.

By early 2026, the warning had evidence behind it. The Stanford

AI Index Report 2026 documented that U.S. software developers aged 22 to 25 had seen employment fall nearly 20 percent from 2024, even as the headcount for older developers continued to grow. Across the most AI-exposed occupations, workers aged 22 to 25 saw employment drop roughly 16 percent relative to the least-exposed jobs, a gap that began widening in mid-2024 and has grown steadily since. Researchers have taken to calling these junior workers the canaries in the coal mine and have named the pattern seniority-biased technological change: AI substitutes for junior labor while leaving senior roles intact. The missing middle has moved from prediction to measured outcome.

This is the industrial bargain playing out in reverse. Taylorism extracted short-term productivity by suppressing the individual, and the over-automation of knowledge work may extract short-term efficiency by undermining the development of expertise. The mechanisms differ. The structural flaw is the same: engineering short-term gains in ways that undermine the future.

The bargain Taylor built rewarded humans for behaving like machines, and Mann's schoolhouse trained children to accept that bargain before they ever had a job. Together they produced a workforce remarkably well suited to a century of industrial production, and remarkably ill-prepared for the moment when the machines learned to do the machine work themselves. That moment is here. The next chapter examines what these machines are, and what they cannot do, before we ask what the human side of the renegotiation should look like.

CHAPTER 4 NOTES

Taylor, Frederick Winslow. The Principles of Scientific Management. Harper & Brothers, 1911. The full sentence reads: "In the past the man has been first; in the future the system must be first." Taylor's work set the intellectual foundation for industrial management through the twentieth century.

Mann, Horace. Seventh Annual Report of the Secretary of the Board of Education. 1844. Mann's enthusiasm for the Prussian system has been analyzed by several educational historians. See Unger, Harlow Giles. Horace Mann: A Biography. John Wiley & Sons, 2012. See also Gatto, John Taylor. The Underground History of American Education. Oxford Village Press, 2001, for a critical interpretation of the compliance logic later associated with the model.

Land, George, and Beth Jarman. Breakpoint and Beyond: Mastering the Future Today. HarperCollins, 1992. Land's longitudinal creativity study tested 1,600 children at ages five, ten, and fifteen, with an adult comparison sample. The instruments were originally designed for NASA engineer selection. The study is often cited as evidence of a decline in divergent-thinking scores with age, but it does not by itself prove a single cause for the decline.

Herzberg, Frederick. "One More Time: How Do You Motivate Employees?" Harvard Business Review, January-February 1968. Herzberg's Hygiene-Motivator theory distinguishes between factors that prevent dissatisfaction and factors that produce genuine motivation. For the original empirical basis, see Herzberg, Frederick, Bernard Mausner, and Barbara Bloch Snyderman. The Motivation to Work. John Wiley & Sons, 1959, based on interviews with 203 engineers and accountants in the Pittsburgh area.

Autor, David H., and David Dorn. "The Growth of Low-Skill Service Jobs and the Polarization of the US Labor Market." American Economic Review 103, no. 5 (2013): 1553-1597. For Autor's specific analysis of AI and expertise commodification, see "Applying AI to Rebuild Middle Class Jobs," MIT Work of the Future, 2024.

Microsoft Research. New Future of Work Report 2025. Microsoft Corporation, December 2025. The report's section on "tacit skill erosion" documents specific concerns about automated entry-level work reducing the development pipeline for expertise.

Gallup. State of the Global Workplace 2023 Report. Gallup, Inc., 2023.

Stanford University Human-Centered Artificial Intelligence. AI Index Report 2026. Stanford HAI, April 2026. The 2026 Index reports on age-stratified employment effects in AI-exposed occupations, drawing on US Bureau of Labor Statistics data through Q4 2025.

CHAPTER 5

From Prediction to Action

The first time you interact with a sophisticated AI system, the sensation can feel like encountering something new. You type a prompt, the cursor blinks, and then comes a response that's coherent, contextually aware, often surprisingly useful. There's a temptation to anthropomorphize immediately, to project intent, wisdom, even understanding.

The pattern is consistent across the educators and executives I work with. We reach for the most familiar frame available and assume something is "thinking," or at least reasoning. The frame is the wrong one, and reaching for it produces a long string of small errors that compound into bad policy, poor pedagogy, and shoddy work.

Before we can think clearly about what AI reveals about us, we need to think clearly about what AI is. And in 2026, that requires updating the picture significantly.

I used to explain it this way: AI is a prediction machine. Ask it for a response, and it calculates which words are most likely to follow, based on patterns learned from vast datasets of human writing. Technically, that's still the base layer. The base layer no longer tells the whole story.

By early 2026, frontier models could complete hour-long software engineering tasks, score in the upper percentiles on professional licensing exams, and run sustained agentic workflows across multiple steps before requesting human review.[1] The capability frontier is a moving target, and the arguments in this book do not depend on which lab leads at the moment of reading. The shift that matters is qualitative. AI no longer only generates text. It takes actions, uses tools, and coordinates sequences of steps toward a goal, and it interprets images, code, and documents alongside language. The result is a class of system that occupies a different role in the workflow than the chatbot of 2 years ago.

The shift from prediction to action has practical consequences. When AI could only generate text, the human remained in the loop by default: you had to read the output, evaluate it, and decide what to do with it. When AI can take actions in the world, searching the web, writing and executing code, sending emails, updating databases, the nature of human oversight changes entirely. By early 2026, multi-agent systems were in production at several major technology companies. When a network of agents coordinates to research a topic, draft a report, format it, and send notifications to the people involved, the human review points become fewer and further between. Accountability, the "who signs this?" issue, becomes more complex when decisions are distributed across a chain of automated steps, and the human who approved the final output may have had no meaningful visibility into the earlier stages.

The right response is precision. Surpassing human accuracy on a defined task is a different thing from surpassing human thinking across all the contexts that surround that task. A model that fills in a standard form more accurately than a human does not thereby understand when the form is the wrong tool for the situation.

What the Systems Still Cannot Do

In 1980, the philosopher John Searle described a thought experiment. A person is locked in a room. They don't speak Chinese, but they have a rulebook in English that tells them, when they see one symbol, to respond with another. People outside the room slip messages in Chinese under the door, and the person follows the rulebook and slips back correct responses. To the outside world, someone in the room speaks fluent Chinese. The person inside has no idea what any of it means.

Searle called this the Chinese Room, and his argument was simple. The room has syntax, the rules of symbol manipulation, but no semantics, the meaning behind the symbols. It can produce the right output without understanding anything about what the output stands for.

In 2026, the Chinese Room is much larger. It can search the internet, run code, and reason through multi-step problems, and the philosophical point holds anyway. Syntax, no matter how sophisticated, does not become semantics.

AI is an extraordinary mapmaker. It has charted the entire territory of human language and much of human knowledge, and it has never touched the territory of human experience.

Uncertainty about a diagnosis means something different for a medical professional. It carries emotional weight. It motivates a kind of careful, recursive attention: checking assumptions, reconsidering the history, asking the patient about something that seemed minor. A physician in that state is doing something that cannot be replicated by a system that generates its outputs without experiencing anything.

When a language model produces an uncertain response, it doesn't experience uncertainty. It produces tokens that represent uncertainty. There is no one inside the room who is worried, no one whose thinking is sharpened by the stakes. The system generates the probability distribution over possible outputs with the same equanimity whether the stakes are high or low.

The Stochastic Parrot

In 2021, the linguists Emily Bender and Timnit Gebru, with their colleagues Angelina McMillan-Major and Margaret Mitchell, published a paper that became one of the most-cited critiques of large language models. They proposed a metaphor that has only grown more useful with time: the Stochastic Parrot.

A parrot can mimic the sounds of human language with unsettling accuracy. It can say "I love you" with flawless intonation and timing, and there is no concept of love behind the words, no self doing the loving, no other being addressed. What it has is exposure. The parrot has heard the phrase in a context where someone meant it, and the parrot reproduces the sound. The reproduction is impressive. It is also, in the strict sense, empty. There is no one home behind the words.

Bender and Gebru argued that large language models do something structurally similar at vastly larger scale. They are trained on enormous quantities of human language and learn the statistical contours of how words appear next to each other. When prompted, they produce sequences that are locally coherent, often globally coherent, sometimes startlingly sharp, and they produce those sequences without anything underneath them. There is no one in the parrot's head trying to communicate something true, no self for whom the output matters. The parrot is, in the technical term Bender and Gebru chose, stochastic. It samples from probability distributions over learned patterns, and the result mimics meaning while bypassing whatever cognitive process produces meaning in a human speaker.

The metaphor has held up because it names what is missing. When a 2026 frontier model reasons through a complex problem, the reasoning is sophisticated pattern recognition across learned representations, remarkably capable statistical processing applied to the structure of human thought as recorded in language. What it lacks is any experience of the problem. It doesn't know what it feels like to be uncertain, to be wrong and embarrassed, to discover something surprising, to want to get this right because someone else is counting on it. The output reads like the result of those experiences. The experiences themselves never happen.

This matters for delegation. The parrot can repeat the cardiologist's diagnosis. It cannot stand at the bedside and notice that the patient is more frightened than the chart suggests. The parrot can reproduce a teacher's lesson plan; it cannot read the room and adjust the lesson when half the class is lost. We have been training a system that can simulate the products of human meaning-making without making anything itself, and the distinction is small in calm conditions and very large when stakes are high.

Extension and Amputation

Marshall McLuhan taught that every technology is an extension of a human faculty. The wheel extends the foot, the camera extends the eye, the book extends memory across time. These extensions have been the engine of human civilization, allowing what was once the work of a lifetime to become the work of an afternoon. AI extends our memory and our pattern recognition. That is a real miracle, and worth saying out loud. McLuhan also warned that every extension comes with an amputation. The faculty we extend, we tend to atrophy. After we extended our feet with the car, fewer of us walked long distances. After we extended our memory with writing, the tradition of memorizing epic poems disappeared from most cultures. Plato, in the Phaedrus, warned that the technology of writing would weaken human memory, and he was right. The trade-off was generally worth making. The amputation was the cost of the extension, and the extension was so useful that the cost felt small.

What gets amputated when AI extends our ability to synthesize information and reason through structured problems? The productive struggle that builds real understanding, the friction that generates insight, the embodied, effortful process of thinking something through. A medical resident using an AI diagnostic assistant from her first day on the wards may never have to sit with the discomfort of not knowing what is wrong with the patient in front of her. The discomfort is what teaches. The AI removes the discomfort and accepts the trade. Five years later, that resident has been an attending physician for 2 years, and a case arrives where the AI's recommendation is wrong and only the physician's own diagnostic intuition can catch it. The intuition was built, historically, by the discomfort she never had to sit with.

We can prevent the amputation, but it requires conscious resistance to the path of least resistance, and the resistance does not happen automatically. The friction that builds skill is precisely the friction the technology was designed to remove.

In a research project of my own, I generated a dozen 5,000-word science fiction stories to test the boundaries of AI fiction from the inside. The results were diagnostic. AI prose consistently favored generic descriptors over specific sensory detail, stated emotions over embodied reactions, conventional plot beats over earned development, thematic statements over implication through action, balanced sentence structures over varied rhythm, and summary over scene. One sentence from my own analysis caught what I had been seeing across all the drafts: "The story knows what stories sound like, and never gets close enough to its own characters to interrupt the sound." Reading the output felt like watching someone describe a story they remembered. The story itself stayed offstage. The systems have nothing to say, and they can simulate having something to say, which is a different thing from creation.

The Accountability Gap

A 2024 randomized clinical trial led by Ethan Goh and colleagues, published in JAMA Network Open, found that access to a large language model did not significantly improve physicians' diagnostic reasoning compared with conventional resources. A 2025 randomized controlled trial in Nature Medicine found that GPT-4 assistance improved physician performance on open-ended patient-care management reasoning tasks. The American Medical Association reported that 66 percent of physicians surveyed were using health AI in 2024. These are real signals, but task design and human oversight matter.

The physician is still the one who signs the chart, still the one who walks into the room and reads the fear in a patient's face, and still the one who weighs the recommendation against this particular patient's life, values, and circumstances. If the recommendation is wrong, the physician is the one who answers for it.

The AI gives better information faster. The human provides judgment, accountability, and presence. As AI systems become more capable, the gap between information quality and accountability clarity widens.

The Automation Complacency Problem

A specific danger comes with capable, reliable AI systems. We saw the first version with semi-autonomous vehicles. Drivers were told to keep their hands on the wheel and pay attention, but the car drove so well on the highway that drivers checked out, and if the car made a mistake, the human was too disengaged to intervene in time.

The same dynamic appears in cognitive work. When an AI writes 10 emails correctly in a row, you stop reading the eleventh carefully. When the system generates accurate summaries 9 times out of 10, the tenth error slips through. High reliability breeds the kind of trust that disables the vigilance the situation requires.

A 2025 Frontiers in Education study of AI use, metacognitive awareness, and critical thinking in higher education supports a more modest concern. Heavy reliance on AI may weaken habits of independent analysis when students consume outputs passively, and the authors call for more longitudinal research. The study does not show neural atrophy, and it does not measure reduced brain activation. The practical warning remains: cognitive processes stay sharp when we exercise them, and they weaken when we habitually outsource them.

Capable AI systems require more sophisticated human oversight, not less. The human role shifts toward directing the work and auditing what comes back, and that shift demands its own set of skills, which need to be deliberately developed.

What Changes and What Doesn't

The updated picture of reality looks like this. AI can predict, reason, use tools, hold context, coordinate actions, interpret multiple modalities, and complete multi-step tasks. It can augment human capability in medicine, law, engineering, science, and creative work in ways that are real and growing.

What stays the same, under current systems, is this. Whatever one believes about future machine consciousness, no AI system in use today can bear professional accountability, suffer social consequences, or participate in embodied relationship. It cannot sit with uncertainty and know what it means to be wrong about something that matters, cannot generate the provenance that makes trust possible, and cannot choose, in any deep sense, to stand behind what it produces.

Those facts should shape institutional design now. Accountability requires an accountable actor. Trust requires someone with something at stake.

As AI handles more of the execution, the remaining human value concentrates around exactly the things execution cannot provide: judgment, accountability, provenance, standing behind a choice and meaning it.

AI benchmarks, bar exams, medical licensing tests, coding challenges, measure the measurable, not the meaningful. An AI that passes the bar exam knows the law. It doesn't know how to read the emotional dynamics of a deposition, manage a terrified client, or negotiate a settlement by reading the other side's actual priorities beneath their stated positions. The benchmark was always a proxy. AI just makes that obvious.

AI is a map of the recorded past, detailed and getting easier to read. The territory it describes is still ours to live in.

Calibrating Your Oversight

A practical rule for working with AI systems: trust the output in proportion to your ability to catch the error. If you are a lawyer reviewing an AI-generated brief, you know the law well enough to spot a hallucinated citation, and your oversight is calibrated. If you are a marketing manager asking AI to draft a legal disclaimer, you probably do not know enough to catch a confident mistake, and your oversight is decorative. Before you delegate a task to AI, ask yourself whether you could catch a well-formatted error in this domain. If yes, delegate and review. If no, learn enough to review meaningfully, or find someone who can. The most dangerous position is confident ignorance: using AI in a domain you do not understand well enough to audit, and assuming the polished output means the content is sound.

CHAPTER 5 NOTES

[1] As of April 2026, capability snapshot drawn from publicly reported benchmarks and product releases. See Google DeepMind, "Gemini 2.5: Our most intelligent AI model," March 25, 2025; Anthropic, "Introducing Claude Opus 4.6," February 5, 2026; and UK AI Security Institute, Frontier AI Trends Report, December 2025. AISI reported that leading systems could complete hour-long software engineering tasks

more than 40 percent of the time, compared with under 5 percent two years earlier, and that the duration of some unsupervised cyber tasks was roughly doubling every eight months. Specific model names and benchmark scores are useful for dating purposes and likely to be superseded by the time this book reaches readers.

Searle, John R. "Minds, Brains, and Programs." Behavioral and Brain Sciences 3, no. 3 (1980): 417-457. The Chinese Room argument remains one of the most discussed thought experiments in philosophy of mind.

Bender, Emily M., Timnit Gebru, Angelina McMillan-Major, and Shmargaret Shmitchell. "On the Dangers of Stochastic Parrots: Can Language Models Be Too Big?" Proceedings of the 2021 ACM Conference on Fairness, Accountability, and Transparency (FAccT '21), 2021. Margaret Mitchell published under the pseudonym Shmargaret Shmitchell.

McLuhan, Marshall. Understanding Media: The Extensions of Man. McGraw-Hill, 1964. Plato's warning about writing and memory appears in the Phaedrus, 274c-275b.

Goh, Ethan, et al. "Large Language Model Influence on Diagnostic Reasoning: A Randomized Clinical Trial." JAMA Network Open, 2024. doi:10.1001/jamanetworkopen.2024.40969. See also Goh, Ethan, et al. "GPT-4 Assistance for Improvement of Physician Performance on Patient Care Tasks: A Randomized Controlled Trial." Nature Medicine 31 (2025): 1233-1238.

American Medical Association. "Physician enthusiasm grows for health care AI." Press release, February 12, 2025. The AMA reported that 66% of surveyed physicians used health AI in 2024, up from 38% in 2023.

National Transportation Safety Board. Collision Between a Car Operating With Automated Vehicle Control Systems and a Tractor-Semitrailer Truck Near Williston, Florida, May 7, 2016. Highway Accident Report NTSB/HAR-17/02, 2017. See also NTSB. Collision Between Vehicle Controlled by Developmental Automated Driving System and Pedestrian, Tempe, Arizona, March 18, 2018. Highway Accident Report NTSB/HAR-19/03, 2019. These reports document overreliance, driver monitoring, and automation-design failures.

Ninghardjanti, Patni, et al. "Evaluating the impact of AI on the critical thinking skills among the higher education students by combining the TAM model and critical thinking theory." Frontiers in Education 10 (2025): 1719625. The study examines AI adoption, metacognitive awareness, and critical-thinking outcomes. It does not measure neural activation or prove neural atrophy.

For readers interested in the model collapse question, see Shumailov, Ilia, et al. "The Curse of Recursion: Training on Generated Data Makes Models Forget." arXiv 2305.17493, 2023, and Gerstgrasser et al., "Is Model Collapse Inevitable?" arXiv 2404.01413, 2024, which qualifies the original finding when synthetic data is accumulated alongside human data.

CHAPTER 6

The Missing Middle

There is a kind of trustworthiness you can only build through doing the work yourself, even when AI offers to spare you the effort. I learned this writing this book.

I started the way I've started most difficult writing projects, by asking an AI to help me think through the structure.

Within minutes I had an outline: clean logic, a sensible chapter sequence, a structure that covered the territory I wanted to cover. It was useful. It was also, in some way I couldn't immediately name, insufficient.

I sat with that feeling for a few days. What was missing wasn't information or words laid out the correct way, because the outline was accurate. What was missing was the resistance: the process of working out what I actually thought by trying to articulate it badly, failing, trying again, discovering something I hadn't expected. The outline gave me the destination without the journey. And it turned out the journey was where I was going to develop the thinking needed to write the book.

This is the missing middle.

The OECD's Digital Education Outlook 2026 draws a distinction that should be at the center of every conversation about AI in learning. General-purpose AI can improve performance without producing learning gains when the pedagogy around it is weak, while thoughtfully designed educational AI can improve learning. These are different outcomes, and they require different conditions.

The gap between performance and learning is the gap between doing the task and being changed by it.

When AI removes the middle, it removes the site of development.

What the Middle Is

Modern systems are built to move past the middle as quickly as possible. They reward peaks: the launch, the milestone, the result. Digital platforms amplify intensity and emotional extremes because that's what drives engagement, and companies tune themselves for fast outputs under pressure. The quiet work of staying with a difficult problem, of sitting with not-knowing, is undervalued at every level.

Neil Postman argued in Technopoly that technological change is ecological. A new technology changes the whole environment it enters. When you use AI to bypass the middle of a task, you change your relationship with that task. You become a consumer of results, and the process happens without you.

This is a development problem across work and school.

Character develops through mundane moments: choosing to keep going when the novelty wears off, listening when you have nothing to say, sitting with a problem instead of seeking an instant fix. When technology removes the difficulty of the middle, it eliminates opportunities for growth, makes a person reliant on external tools to solve what are, at their core, internal challenges, and trades agency for convenience.

The philosopher Albert Borgmann captured this in his concept of the device paradigm. A fireplace requires you to gather wood, build a fire, tend it, and that activity centers your life and connects you to your environment. A thermostat provides warmth at the push of a button. More efficient, but the engagement is gone, along with the relationship with the material and the skill of reading the fire.

Borgmann was distinguishing between focal practices, activities that engage us meaningfully with our world, and commodity consumption, where we receive outputs without participating in the process. AI, in its most common use, functions as a device. It removes us from the process, and the engagement disappears.

Once engagement goes, the losses run deeper than efficiency. We lose the developmental opportunities built into the difficulty, the errors that teach us things books cannot, and the calibration of our own thinking that comes from repeatedly wrestling with real problems and comparing our predictions to outcomes.

The Expertise Pipeline Problem

There is a technical version of this argument that economists and educators are grappling with directly.

The training ground for expertise is the space between novice and competent: the entry-level work, the routine tasks that require enough knowledge to execute but not enough to make you irreplaceable. That work has historically served 2 functions. It produced value for the organization, and it developed the discernment of the person doing it.

AI is absorbing the first function. It can produce routine outputs faster and more consistently than a junior employee. But the second function requires a human learner, because judgment builds only inside the person doing the work.

The problem under this is structural. If AI handles the junior work, senior people stop getting made, and the next generation of expert discernment has to come from somewhere. Senior expertise is built through years of getting things slightly wrong, noticing, adjusting, internalizing feedback. The learning happens in the doing, and specifically in doing things that are slightly beyond your current capability.

The Microsoft New Future of Work Report 2025 documented this concern directly. Automating too much, especially entry-level work, risks losing the hands-on learning and tacit skills that drive long-term innovation.

Stanford researcher Hari Subramonyam has documented how AI tools designed to produce polished output instead of scaffold learning undermine critical thinking. His research found that when the learner stays passive in shaping the work, the creative and critical decisions are made by the AI. This is partly why so many AI-assisted essays look and sound the same. People are leaving their own ideas and insights out.

The finding runs deeper than writing. It applies to any cognitive skill that develops through engaged practice, whether legal reasoning, medical diagnosis, engineering design, or teaching itself. In each case, the skill builds through working under conditions of genuine uncertainty, not through consuming the outputs of a system that has resolved the uncertainty for you.

The psychologist William James understood this in 1890. In The Principles of Psychology, he wrote that sustained voluntary attention is "the very root of judgment, character, and will." Not information, and not output. Attention, the effortful, directed kind that you have to choose to maintain when your mind wants to wander. James saw this as the foundation of everything we mean by human capability. AI can process attention's outputs. It cannot perform the act of attending.

Anders Ericsson, whose decades of research on expert performance produced the concept of deliberate practice, documented this consistently across music, chess, sports, and medicine. Expert performers don't improve by performing comfortable tasks repeatedly. They improve by targeting skills just beyond their current capability, engaging with real difficulty, receiving feedback, and adjusting. The struggle is the mechanism of learning. A student who cannot yet write a complex argument can have AI produce one, and a consultant who cannot yet synthesize a client situation can have AI draft the synthesis. The output looks like expertise. The underlying cognitive structure has not been built.

Using AI to skip the middle doesn't produce the output of a master and the mind of a master. It produces the output of a master and the mind of a novice. That combination is dangerous. And it scales.

The Struggle That Matters

I spent several days trying to write the introduction to this book. I knew what it needed to cover and had outlined the argument, and still I could not find the way in. I wrote 12 different versions. Each one felt wrong: too abstract, too technical, too casual, too formal, missing something essential I couldn't name.

I could have asked an AI to generate options. It would have produced 10 variations in seconds, polished prose in clear sentences, all of them reasonable.

I didn't do that.

Because the struggle was doing something.

Each failed attempt taught me what I didn't want to say. Each version that felt wrong clarified what I was trying to express. The difficulty forced me to think harder about what mattered in that opening, which tone to set, which problem to raise. By the 13th attempt, I understood something I hadn't understood before I started. The chapter wasn't about what I thought it was about. I had been trying to write an opening for an argument I didn't want to make.

I thought I was experiencing writer's block, but I soon realized it was my mind telling me something was off. The struggle forced me to reconsider the whole chapter, which I ended up restructuring entirely.

That took time. It felt inefficient while I was in it. But the meaning I arrived at emerged from the struggle, not despite it. The difficulty was generative.

A 2024 study by Doshi and Hauser found that writers who used AI-generated ideas produced work that judges rated as individually more creative, but the collection of AI-assisted work was more similar than the collection produced without AI. The homogenization was measurable: collective diversity decreased as individual polish increased. The struggle, it turns out, is where our distinctiveness comes from. My difficulties with this chapter look different from your difficulties with your report, and my weird associations, my specific failures, my particular frustrations produce a voice that is mine.

AI pulls everything toward the center. It gives us the most likely word, the most probable argument, the standard approach, and it smooths away the jagged edges that turn out to be our fingerprints.

The stakes are bigger than aesthetics. Once public communication homogenizes toward the mean, something in public discourse starts to thin. If a large proportion of public communication, journalism, academic writing, policy documents, business correspondence, becomes AI-assisted in the default way, the range of perspectives in circulation narrows. Ideas that diverge from the statistical average are less likely to be generated, and the edges of the thinking space, where most innovation originates, shrink.

The Learning/Performance Distinction

A hard objection belongs here. AI can also create new forms of practice. A student who debates an AI tutor for an hour may learn more than a student staring at a blank page and giving up, and a junior analyst who asks an AI to critique her first draft may learn faster than one waiting a week for feedback. The issue is design. AI becomes dangerous when it substitutes for the learner's effort. It becomes useful when it makes that effort more visible, responsive, and demanding.

The OECD distinction deserves more attention than it usually gets.

Performance is what you produce. Learning is what changes inside you as a result. These often come apart. A student who uses AI to produce an excellent essay may have demonstrated only that AI, properly prompted, can produce excellent essays, while learning nothing about literary analysis, argumentation, or the specific text.

When you had to develop the skill to produce the output, performance and learning tracked together. AI breaks that alignment. A student can produce sophisticated performance without the underlying learning, and a professional can produce competent-looking work without the underlying expertise. The performance happens. The learning underneath it does not.

The consequences are delayed but not avoidable. The student who got through university by leaning on AI arrives in a professional context needing judgment she never developed. The consultant who used AI to produce polished deliverables without doing the underlying thinking faces, eventually, a situation where the AI is wrong and his judgment is needed. And the judgment is missing.

Research cited in the 2025 Frontiers in Education study points to a more cautious claim. Some studies associate heavy or unregulated AI dependence with weaker creative-thinking or critical-thinking outcomes over time. The evidence is early and often correlational, but the warning is worth taking seriously. When students use AI extensively without scaffolding, cognitive flexibility and independent problem-solving may be weakened rather than strengthened.

This is what I would call the deferred developmental cost. You get the performance now and pay the learning debt later, the way a minimum payment hides the size of the balance. And by the time you are paying it, you may not recognize what you are missing, because you never built the baseline that would let you notice the gap.

Reclaiming the Middle

There is unproductive friction, the bureaucratic hoop, the formatting exercise, the repetitive task that teaches nothing, and AI should handle it. There is productive friction, the wrestling with ideas, the difficult conversation, the creative block that forces you to think harder. That friction is the point. It is where development happens. If we let AI remove it, we don't get more time for higher-order thinking. We get a generation of people who never built the ability to do it.

Specific, practical steps matter here.

Write a first draft of important work before consulting AI. The instruction is to do your own thinking first, so that when you consult the AI you can engage with its output critically, comparing it to where you were, and choosing what to keep based on your own sense.

Use AI to challenge your thinking instead of replacing it. Instead of asking AI to generate an argument, give it your argument and ask it to find the flaws, and ask it what the strongest counterargument would be. Use the tool to stress-test your own thinking, not to substitute for it.

Protect the difficult conversations. The temptation to use AI to draft the difficult email, the performance feedback, the conflict resolution message, is strong. But these conversations develop your relational and communicative skills in ways that doing them through a proxy does not. The awkwardness of not knowing exactly the right words is developmental. It builds your ability to handle ambiguous human situations in ways that no amount of AI-polished text can.

Let yourself sit with hard problems before searching for answers. Maintaining attention on a difficult problem, before reaching for a tool to resolve it, builds the tolerance for ambiguity that high-stakes thinking requires, and it develops the kind of patient attention that generates real insight instead of adequate output.

The contemplative traditions have long recognized this quality of quiet, inward attention as the foundation of wisdom, creativity, and genuine self-knowledge. The mystic's silence, the philosopher's study, the artist's studio practice, the scientist's long engagement with a single problem are all versions of the same thing: extended time alone with one's own mind, without distraction or assistance. In a world of constant connectivity and instant answers, this quality of solitude is increasingly rare and more and more essential. The ability to do it, the sense of what you actually think as distinct from what you have absorbed, is one of the developmental products of the missing middle.

The middle is also where the relationship with yourself gets built. Skip it habitually, and we become spectators of our own lives, and spectators can't exercise the kind of discernment that an age of cheap execution demands.

AI gives us a real choice. We can use the time it saves to consume more, to push through more tasks, to produce more output with less engagement. Or we can use the time it saves to go deeper: to have the difficult conversation, read the long book, sit with the hard problem, and develop the discernment the next task will require.

The argument for protecting the middle is pedagogical and neurological, and at its deepest, it concerns what kinds of knowing we are willing to cultivate.

The middle rarely feels like progress and almost never looks productive from the outside. It is where the self gets formed, expertise gets built, and meaning gets made. AI can help us protect that middle, but only if we understand why it matters before we start smoothing it away.

CHAPTER 6 NOTES

Organisation for Economic Co-operation and Development. OECD Digital Education Outlook 2026. OECD Publishing, 2026. Available at oecd.org/en/publications/oecd-digital-education-outlook-2026.

Postman, Neil. Technopoly: The Surrender of Culture to Technology. Knopf, 1992.

Borgmann, Albert. Technology and the Character of Contemporary Life: A Philosophical Inquiry. University of Chicago Press, 1984.

Microsoft Research. New Future of Work Report 2025. December 2025. Section 3: "Implications for Workforce Development."

Stanford Graduate School of Education. "Beyond ChatGPT: Critical Thinking in the Age of AI." GSE News, 2025. Quoting Hari Subramonyam, assistant professor at Stanford GSE.

Doshi, Anil R., and Oliver P. Hauser. "Generative AI Enhances Individual Creativity But Reduces the Collective Diversity of Novel Content." Science Advances 10, no. 28 (2024): eadn5290.

Ninghardjanti, Patni, et al. "Evaluating the impact of AI on the critical thinking skills among the higher education students by combining the TAM model and critical thinking theory." Frontiers in Education 10 (2025): 1719625. See also Lee, H. P. H., et al. "The Impact of Generative AI on Critical Thinking." Microsoft Research, 2025, for survey evidence on cognitive offloading in knowledge work.

Ericsson, K. Anders, and Robert Pool. Peak: Secrets from the New Science of Expertise. Houghton Mifflin Harcourt, 2016.

PART III

The Human Advantage

CHAPTER 7

The Context Advantage

For more than 150 years, we ran a massive unintentional experiment on the human mind. We turned our education systems, corporate training programs, and cultural values into content delivery machines, tuned everything for information transfer, and measured intelligence by how many facts you could store and how fast you could retrieve them. We told our children that the humanities were a luxury, a degree for the unemployable.

On November 30, 2022, the experiment reached its conclusion.

We built a machine that could store more content, retrieve it faster, and synthesize it more comprehensively than any human being who has ever lived. If the goal of education is to answer "What is the capital of France?" or "How do I write a Python script for a bubble sort?", humans are no longer the best tool for the job.

This has caused panic, because the skills we thought would last forever are turning out to be the most easily automated.

At the same time, something strange is happening to what we used to call soft skills. Understanding history, constructing a logical argument, reading between the lines, weighing the ethical implications of a decision, communicating across disciplines: these are suddenly becoming the rarest and most prized capacities available. The panic about content is obscuring the central shift. When content becomes abundant, context becomes the scarce capability.

Content and Context

Content is the "what": the data, the facts, the words on the page, the pixels on the screen. AI is an extraordinary content engine. It can generate a million essays on Hamlet in an hour or produce a thousand marketing strategies in a minute.

Context is the "why": the web of meaning surrounding the content, the understanding of who is speaking, when they're speaking, what they're trying to achieve, what the situation requires, and what a particular human being in this particular moment needs.

A medical AI can look at a list of symptoms and generate a differential diagnosis. Only a human doctor can look at the patient, see the fear in their eyes, know they just lost their spouse, and recognize that the chest pain is probably grief, not cardiac.

An AI legal assistant can find every precedent for a contract dispute. Only a human lawyer can read the situation and understand that the client doesn't want a win so much as an apology that repairs a relationship.

In each case, the machine provides the text. The human provides the context. Compete with the machine on content, and you will lose. Become the context engine, and you become the part the machine cannot replace.

The ancient Greeks had 2 words for knowledge: episteme and phronesis. Episteme is theoretical knowledge, the kind stored in books and databases. Phronesis is practical wisdom, the kind built through experience and applied to specific situations. For centuries, the education of professionals focused primarily on episteme, because episteme was hard to acquire and distribute. AI has effectively commodified episteme, making it abundant and cheap, while phronesis remains scarce. It requires time, experience, and engagement with reality that no system can simulate.

The context advantage is a phronesis advantage. And it grows as episteme becomes more abundant.

The 2025 Labor Market Signal

The creative economy research group PEC published a study in late 2025 analyzing UK job vacancies after ChatGPT's arrival. They found that demand for creativity had not declined but had grown more sophisticated: the job postings requiring creative skills were increasingly bundling them with domain expertise and discernment, rather than asking for isolated creative ability.

The market was asking for contextually grounded creativity. The ability to generate was now everywhere. The ability to make it matter was rare.

PwC's 2025 Global AI Jobs Barometer documented that skill demands are changing 66 percent faster in AI-exposed jobs than in jobs with less AI exposure, and the skills with growing demand are contextual: leadership and team management, client relationship building, creative problem-solving, strategic thinking and planning. These are all forms of applied phronesis, judgment exercised in specific relational and organizational contexts.

The pattern is visible in the productivity data as well. The 2026 AI Index summarized several micro-level studies showing AI productivity gains of 14 to 26 percent on structured, language-heavy tasks such as customer support, software pull requests, and ad creation. On tasks requiring more judgment, the same studies report weaker, zero, or negative effects. One 2025 study found that experienced software developers using AI for open-source work were about 19 percent slower than developers working without it, even though the developers themselves believed they were going faster. The tools speed up what can be templated. They do not speed up, and may impede, what requires situational reading.

The McKinsey analysis found that occupations with the strongest demand growth through 2030 share a common feature: they involve judgment under conditions of uncertainty in high-stakes domains, in fields like healthcare, law, education, complex engineering, and creative direction. These are fields where the context is as important as the content, where the right answer depends on who is involved, what they need, and what values are at stake.

The New Competence Stack

This realization changes what we should be developing in people, and in ourselves. The new competence stack cuts across humanities and STEM, because the old binary is too coarse. What the context advantage demands is a specific set of integrated capacities.

Domain knowledge. You need to understand something well enough to audit AI outputs in that area. Whether it's medicine, law, finance, engineering, or education, you need the technical grounding to know when the machine is wrong or incomplete. The generalist who uses AI without domain expertise cannot tell when the output is confidently incorrect.

Rhetoric. The Greeks understood this. Rhetoric is the art of using language to achieve a specific result with a specific audience, and prompt engineering is rhetoric applied to AI. George Orwell argued that clarity of language reflects clarity of thought. The person who asks AI to write "in a Strunk and White style, avoiding adverbs and prioritizing active verbs" will get a better result than the person who says "professional tone," and that difference comes from having read, practiced, and internalized the craft of language.

Source judgment. Evaluating the credibility, provenance, methodology, and limitations of information. In a world where AI can generate plausible-sounding content on any topic, distinguishing reliable from unreliable, primary from secondary, verified from generated becomes essential. This is what historians, journalists, and scientists develop through training, and it transfers.

Add to this statistical literacy (reading a chart critically and spotting correlation masquerading as causation), ethical reasoning, and organizational sense, the reading of how institutions work.

Taste. Knowing what is good before the data proves it, and looking at 10 AI-generated options and identifying the one that will resonate with a human being. Taste is discernment applied to quality. It develops through wide exposure, careful attention, and deliberate practice, and it cannot be automated.

These capacities form judgment under uncertainty. They integrate information, values, context, and responsibility into a choice you can defend. The humanities, broadly understood, develop this kind of judgment. Knowledge of history, literature, and philosophy matters here for what it builds: the intellectual toolkit that produces the judgment this moment requires.

The Medium Is the Message, Again

Becoming the context engine, after a lifetime of being rewarded as a content creator, requires understanding AI as a medium. Marshall McLuhan taught that the medium is the message. The tools we use to communicate shape our ideas while carrying them, and they also shape the institutions, relationships, and ways of seeing that surround them.

The printing press did more than put ink on paper. It created the conditions for the Reformation, the Enlightenment, and the modern concept of intellectual property, but even that revolution took time to be understood. The first users of the press merely reproduced religious manuscripts faster, and it took decades before anyone imagined scientific journals, novels, or mass education. When electricity arrived in factories, manufacturers spent 30 years replacing steam engines one-for-one before recognizing the potential to redesign entire production systems.

New technologies demand imaginative reconfiguration, not mechanical substitution. AI is no different. The organizations using it only to automate existing workflows are in the steam-engine-replacement phase. The real shift requires the context to imagine something new.

Every medium demands a price. The printing press cost us oral memory. AI may cost us cognitive friction, the struggle replaced by speed and smoothness, the "most likely" answer arriving before the hard issue has been fully formed.

Without grounding in critical thinking, without examining what a tool is doing to us, we risk becoming passive. Context advantage depends partly on stepping back and asking what this is doing to how I see, what it is making easier that should not be easy, and what it is making invisible that should be visible.

McLuhan also observed that every technology retrieves an older form when pushed to its limit. AI, pushed to its limit, may retrieve exactly the things it appears to replace: rare, slow, embodied, personally voiced human expression that only someone with lived experience can provide. The content is cheap. The context is everything. And context always comes from somewhere, from the specific history, the specific perspective, the specific values of a person who has been somewhere and seen something. That comes only from lived experience.

The legal industry is already producing the interpreter role. The law firm that uses AI for document review and research still needs human lawyers who can evaluate whether the AI's analysis holds up against the specific facts of the case, whether it has missed the particular judge's interpretive tendencies, whether the client's actual situation is being properly captured. That human layer is where the value lives.

Medical practice is producing it too. The hospital system that uses AI for diagnostic imaging still needs clinicians who can integrate the AI's findings with everything else they know about the patient, including the conversation in the exam room, the subtle presentation cues that didn't make it into the structured data, and the patient's values and preferences about treatment. The AI processes the image. The clinician interprets the person.

The Return of Rhetoric

AI forces a return to something the ancient world took seriously: the art of the prompt.

Cicero's teachings on rhetoric emphasized the rhythm of language, the structure of argument, the importance of audience. These skills felt antiquated in a world of emojis and 280-character posts. They are suddenly relevant again.

When you interact with a language model, you're engaged in a rhetorical act. The machine is remarkably literal. It doesn't understand "vibes." It responds to precise instructions. If your thinking is muddy, your prompt will be muddy. If your thinking is clear, if you know what you want and can describe it precisely, the machine will perform accordingly.

This extends beyond prompting. The person who can articulate a complex situation clearly to a board of directors, a patient, a team, or a student has a skill that AI doesn't replicate. AI can produce technically correct explanations. Human articulation requires reading the person, adjusting in real time, choosing what to include and what to leave out based on what the situation requires.

That's rhetoric. And the market for it is growing.

The What-For

We've spent decades rewarding the "how": how to build the bridge, how to code the app, how to structure the tax shelter. AI is also expert at the how. When the how gets cheap, the what-for becomes the work.

What is this website actually for? Whether it serves its users or extracts their attention for someone else's benefit, whether the contract protects a partnership or merely armors one side, what the data means for a community's future: these are the questions the how can't answer. They are philosophical concerns, and sociological and psychological ones too, and they require a kind of knowledge that develops through engagement with exactly the disciplines we've been devaluing.

The marketer who uses AI to analyze customer data, then applies knowledge of anthropology to understand the culture behind the data, can recognize that a decline in sales is about a shift in tribal identity that the algorithm missed. The developer who uses AI to write code, then asks "who gets left behind if we release this algorithm?", builds products that work without harming. The teacher who uses AI to grade grammar, then notices that a student's essays have become dark and withdrawn, can discover that the student is in crisis, something no grammar checker would flag.

In each case, the human adds a layer of meaning the machine cannot access. Wisdom. Intelligence wrapped in context, the knowing of what to do when the rules don't apply.

The Rise of the Interpreter

In the age of abundant data and automated analysis, the scarcest cognitive service is interpretation.

Anyone with access to AI tools can now produce data analysis. The data is abundant, the tools to process it are accessible and getting cheaper, and meaning has become the bottleneck. What does this data mean for us, in our context, given our values, our history, our specific circumstances?

This requires an interpreter: someone who understands the domain well enough to evaluate the analysis critically, knows the organization's history and culture well enough to understand what the data is measuring and what it's missing, can read the political and relational dynamics the data doesn't capture, and can articulate the implications clearly enough to drive decisions.

The interpreter goes beyond the analyst. The analyst processes data according to defined methods, while the interpreter makes the calls about which methods are appropriate, what the results mean, and what should be done about them, bringing domain knowledge, organizational context, and human sense to the output of the analyst's work.

In most organizations today, the analyst and the interpreter are different people in different roles. As AI absorbs more of the analyst function, the interpreter role becomes more central, and the ability to make sense of AI-generated analysis, challenge its assumptions, and situate its findings in the broader context of what the organization needs becomes one of the genuinely scarce capabilities in the AI economy.

The interpreter is also the translator, the person who can operate across the boundary between technical and humanistic, between what AI does and what it means for people. The data scientist whose models produce detailed risk assessments needs someone who can communicate the implications to a board that doesn't speak statistics, and the AI system generating medical recommendations needs a clinician who can make those recommendations actionable for a specific patient with a specific cultural background and emotional state. Every field using AI widely needs people who can bridge that gap. Translation here is a human act. What does this mean for this person, in this situation, with these values and these needs?

The context advantage comes from a specific combination: enough technical grounding to understand what the AI is generating, enough domain expertise to evaluate it critically, and enough contextual intelligence to translate it into action appropriate for the specific situation at hand.

There is a particular irony worth naming. For more than 2 decades, the dominant narrative held that the future belongs to STEM and the humanities are a luxury. AI has reversed this calculation. The technical skills that commanded premium wages in 2015 are now the most readily automated, while the skills that command premium wages now are the ones AI cannot replicate: contextual judgment, relational intelligence, ethical reasoning, communicative sophistication. The humanities are coming back, and not by accident. The capacities belong to anyone who has built them, with or without the formal degree.

Interpreting AI output calls for a developmental orientation: the ability to ask better prompts, because AI is only as good as the guidance it receives; a tolerance for sitting with uncertainty, because context is always incomplete; the ability to reason ethically, because context includes values; and a commitment to communication, because the interpreter's value is only realized when the interpretation is transmitted in a form that others can receive and act on.

These are old skills that are becoming newly scarce. Schools that develop these capacities are recovering something that got lost in the long run of measuring the measurable, grading the gradeable, standardizing the standardizable. The interpreter is what a liberally educated professional was always supposed to be. AI's arrival is making that aspiration urgent in a way it wasn't before.

The organizations that will perform best over the next decade will win on something besides the best AI tools, because every organization will have access to roughly comparable tools. The winners will be the ones who develop the most capable interpreters, the people who can read context, translate data into insight, and make decisions that are right for their specific situation. That advantage is durable in a way tool advantage can never be.

Building Your Context Stack

Look at the 7 parts of the competence stack and pick the weakest one. Not the one that sounds most impressive: the one where you have the biggest gap between what you know and what the moment requires. That is where your edge is. Spend the next month developing it. Read 1 book in that area, have 3 conversations with someone who has the skill you lack, and practice it on 1 real problem at work. Domain knowledge develops through immersion. Rhetoric develops through writing and speaking with care. Source judgment develops through reading critically, statistical literacy through pressure-testing the numbers, ethical reasoning through sitting with hard cases, organizational sense through watching how decisions get made, and taste through wide exposure and honest reaction. None of these are abstract, and all of them are buildable. Start with one.

CHAPTER 7 NOTES

Stanford HAI. AI Index Report 2026. Institute for Human-Centered AI, Stanford University, April 2026. Chapter 4, "Economy," documents the widening employment gap for workers aged 22 to 25 in AI-exposed occupations, drawing on studies from Brynjolfsson et al. (2025), Hosseini Maasoum and Lichtinger (2025), and the Federal Reserve Bank of St. Louis (2025). The "canaries in the coal mine" and "seniority-biased technological change" framings come from these studies as cited in the report.

Wang, Z., Baksy, A., Bakhshi, H., and Siepel, J. "Demand for Creativity and AI Skills in the Post-ChatGPT Labour Market: Evidence from UK Job Vacancies." Creative PEC Discussion Papers Series, November 2025. Available at pec.ac.uk.

PwC. The Fearless Future: 2025 Global AI Jobs Barometer. PricewaterhouseCoopers, 2025. The finding that skill demands are changing 66% faster in AI-exposed jobs comes from longitudinal analysis of job posting requirements.

McKinsey Global Institute. "Agents, Robots, and Us: Skill Partnerships in the Age of AI." McKinsey & Company, November 2025. The demand growth analysis examines occupation categories with sustained employment expansion through 2030 projections.

Orwell, George. "Politics and the English Language." Horizon, April 1946. Orwell's essay argues that vague or imprecise writing is a symptom and a cause of muddled thinking, and that clarity of expression is both a political and intellectual virtue.

McLuhan, Marshall. Understanding Media: The Extensions of Man. McGraw-Hill, 1964. The phrase "the medium is the message" refers to McLuhan's central thesis that the characteristics of a medium shape the kinds of messages it transmits and the culture that uses it, independent of the specific content being transmitted.

U.S. Copyright Office. Copyright and Artificial Intelligence, Part 1: Digital Replicas. Library of Congress, July 31, 2024. See also U.S. Copyright Office. Copyright and Artificial Intelligence, Part 2: Copyrightability. Library of Congress, January 29, 2025.

For research on customer service and AI vs. Human interaction outcomes, see Luo, Xueming, et al. "Frontiers: Machines vs. Humans: The Impact of Artificial Intelligence Chatbot Disclosure on Customer Purchases." Marketing Science 38, no. 6 (2019): 937-947.

CHAPTER 8

The Practice of Presence

It was a Tuesday morning in March 2025. I was at my desk, in the small home office where most of this book got written. Three monitors. Coffee gone cold. An AI conversation open in one window, a research document in another, an email I had been meaning to answer for 2 days in a third. I was working on what would become chapter 6.

My wife came to the door of the office. She said something. I answered without looking up. She said something else. I made a noise that could have meant agreement or acknowledgment. She left.

A few minutes later I realized I did not know what she had said. Not the words, not whether she needed an answer, not whether she was happy or worried or just passing through. The exchange had happened in front of me, and I had been somewhere else. Absent, though the body stayed at the keyboard, in the more important sense, the one the book had been arguing about for several months by then.

I sat with that for a minute. Then I closed the laptop.

I am writing a book about how AI reveals what we have lost, and the morning I most needed to see what the book had been telling me, I missed it because I was inside the technology that was the subject of the book. The mirror went up. I was not looking.

This chapter is about the discipline that makes the looking possible.

What Presence Is

The word presence has been thinned out by use. It now means something close to attentiveness: eye contact, putting your phone down at dinner. These are good things. They are not what the contemplative traditions have meant by presence for more than 2,000 years.

Presence in the older sense is a trained skill, the ability to remain inside an experience, including the uncomfortable parts, without immediately moving to fix it, label it, or escape into the next thing. The traditions have called it many things: hesychia in the Greek-speaking East, recollection in the Latin West, mindfulness in the Pali sources, wakefulness in some Zen lineages. The vocabulary differs. The disciplined attention they describe is recognizably the same thing.

Real presence is often uncomfortable, especially at first. Thoughts continue to arise inside it. Feelings come and go. The discipline is the choice, repeated thousands of times in a life, to stop running and pay attention to what is here.

That choice is harder than it sounds. The cultures most readers of this book live inside have spent the better part of 2 centuries training the opposite habit, the habit of scanning ahead, anticipating the next thing, tuning the present moment for some downstream payoff. Presence asks for a different orientation. It asks for the present moment to be enough.

Four Modes of Mind

The British psychologist John Teasdale, who spent decades researching the cognitive mechanisms of mindfulness-based therapy, named 4 modes of mind that map onto everyday experience. They are useful here because they describe with some precision the difference between what knowledge work has been training us to do and what presence requires.

The first mode is Driven-Doing. This is the mode of most modern professional work. The mind scans constantly for the gap between the current state and the desired state, then reaches for whatever will close the gap. Email arrives, and anxiety rises until it is answered. A meeting is scheduled, and tension builds until it is over. A deadline approaches, and energy spikes until the deliverable is shipped. Driven-Doing exhausts the person who lives in it. It is also, by its nature, never satisfied. The instant one gap closes, the mind locates the next one.

The second is Doing. Same task focus, less reactive urgency. You are working steadily on something because the work matters and you have decided to give it your attention. The discrepancy-scanning is quieter. You are not running from one finished thing to the next.

The third is Being. The mind has no task at all. It is taking in what is here: walking the dog without listening to a podcast, sitting with a cup of coffee and not also reading. The body relaxes. Awareness opens. Most adults in most working weeks visit Being for very little of the day, and often only when they are tired enough to stop chasing.

The fourth is Mindful Being: the same quality of open awareness as Being, but now sustained on purpose, by intention. You are not waiting for the activity that would put you back into Driven-Doing. You are practicing the ability to remain in the present moment without immediately leaving it.

Teasdale's frame helps because it names the problem. Knowledge work, even at its most thoughtful, lives mostly in Driven-Doing, and AI does not move us into one of the other modes. AI speeds up Driven-Doing. The cycle from gap-detection to gap-closure shortens, the number of gaps available to detect rises, and the mind stays in the same mode it has been in for the last several decades, only faster.

If the argument of this book is that AI's arrival has clarified what human capability is for, the answer is going to involve modes of mind that AI cannot speed up. Being and Mindful Being are where the work has to happen.

The Watered-Down Version

Mindfulness, as it has spread into corporate wellness programs and meditation apps and stress-reduction modules, is mostly a softer version of Driven-Doing dressed up as something else. I want to say that carefully, because I know how it sounds.

There is nothing wrong with a 10-minute meditation app. The teachers building those products have helped a lot of people, and the watered-down version is a real entry point. The problem comes next.

The contemplative traditions did not propose meditation as a stress reduction technique that would make you a better executive. They proposed it as a complete reorientation of how you live. The practice was a discipline embedded in a community, a theology, a daily rhythm, and a willingness to let the practice slowly disassemble the self that had been holding everything together.

The 10-minute app version takes the technique out of that frame and offers it as a productivity supplement: Headspace before the meeting, Calm before bed, track the streak. The practice gets pulled into Driven-Doing instead of escaping it. You meditate to perform better at the thing you were already doing too much of.

I have a friend who told me, with no irony, that he had stopped meditating because he kept getting frustrated with his progress. He had been measuring his streak in the app for almost a year. The mind that wanted to win at meditation was the same mind that needed meditating. The app had not changed his mind. It had given his mind a new domain to dominate.

The apps have their place. What this moment requires is the deeper version of the practice, the version that cannot be measured, polished, or completed, the version that asks you to be a different kind of person rather than a more efficient version of the person you already are.

What the Contemplatives Knew

Across the contemplative traditions, the disciplines converge on a few elements. Silence. Stillness. A return of attention to a single point, repeatedly, without expectation that the return itself will produce a payoff. A willingness to encounter what is rather than what should be, and a trust that something happens in the encountering that cannot be summoned by effort.

The Franciscan teacher Richard Rohr, whose work has shaped my thinking more than I could fairly summarize in a paragraph, distinguishes between dualistic and non-dual consciousness. Dualistic mind sorts the world into categories: right and wrong, mine and yours, useful and useless. It is fast, and it is necessary for some kinds of decisions. It is also the operating mode of most public discourse, most institutional life, and most of the internal commentary running through any given person's day. Non-dual consciousness can hold what dualistic mind cannot. The friend who is also disappointing me. The work that is necessary and partly meaningless. The grief that arrives in the middle of a celebration. The teacher who is wrong about some things and life-changing about others. Rohr's argument, drawn from a long tradition, is that the practice is the slow, often painful work of training non-dual awareness in a culture that rewards the dualistic kind.

This matters for AI because AI is, in a precise sense, the most powerful dualistic mind we have ever built. It sorts, categorizes, and pattern-matches, producing confident outputs by reducing ambiguity. Non-dual is beyond it. It cannot remain with contradiction without trying to resolve it.

Much of human meaning lives inside contradiction, and the

places where presence is required are the places where the contradiction must be held, not solved. A bedside conversation. A failing marriage that might still be worth saving. A team member whose work is slipping for reasons you do not yet understand. A creative problem with no clean answer. The dualistic mind, including the artificial dualistic mind, is the wrong tool for that work.

The contemplative practice slowly grows the right tool.

Where I Fail

I want to put my own honest experience on the page here, because the book is about doing the thing instead of just arguing for it.

I have been around this material for almost 20 years. I have read the texts. I have led conversations with educators and executives where I made the case for presence with what I hope was genuine conviction. I still default to Driven-Doing every morning before I am awake enough to know better.

The day I described at the beginning of this chapter was not unusual. It was Tuesday. There have been hundreds of Tuesdays. I close laptops. I open them again. I walk into the kitchen and immediately reach for the phone. I notice the reach. I do it anyway. I read a contemplative book about presence while my mind is composing the email I will send when I finish the page.

The Enneagram lineages would say I am a Seven: a reframer, an avoider of difficult feelings, a person whose default is to keep moving. The label is useful only in the sense that it names the pattern. It does not solve it. The practice solves it, slowly, and only sometimes.

The reason I am writing this chapter is that the practice has changed what is possible for me, even though I fail at it most days. The wife I missed in March 2025 is still my wife. The marriage is still a marriage. The work is still getting written. None of that is automatic. All of it depends on a discipline I am barely competent at, that I keep returning to, that I have never finished.

The honest sentence: presence is a place you return to, never a state you achieve and keep.

What the argument of this chapter asks of the writer is willingness to keep returning to the discipline. Skill at it is optional. That is the only credential the contemplative tradition asks for. Show up. Begin again.

The Body in the Room

Iain McGilchrist's hemispheric argument is worth bringing in here, because presence is, at the level of cognitive architecture, a return of attention to capacities the dominant culture has trained us to suppress.

The right hemisphere, in McGilchrist's account, is the seat of embodied, contextual, whole-field awareness: the here and now, the body in the room, the face of the person across the table, the unspoken cue, the felt sense that something is wrong before the analytic mind has named what. The left hemisphere is the seat of categorization, language, abstraction, and goal-directed action. Each mode is necessary. The book argues that Western culture, since the printing press, and especially since the industrial revolution, has trained us to live mostly in the left hemisphere's strengths and to mistrust the right's.

AI lives in the left hemisphere's domain. It is excellent at categorization, language, abstraction, and goal-directed action, and it has no body, no here, no felt sense. The right-hemisphere capacities are, structurally, the human inheritance AI cannot occupy.

Presence is, in part, the practice of returning attention to the body in the room: the breath, the temperature of the air, the expression on the face of the person who just walked in. Details like these are the data the analytic mind cannot fully access from text, and they are also the data on which most consequential human judgment depends. The clinician who senses the patient is more frightened than the chart suggests is reading right-hemisphere information. The leader who can tell the team is exhausted before anyone says it is doing the same.

Knowledge workers have been trained for several decades to ignore right-hemisphere information in favor of dashboard information. AI speeds up that training. The presence practice is partly a deliberate counter-current.

The Loss of Solitude

Solitude has a specific role in the contemplative traditions. It is a central condition under which presence becomes practicable, and most of the deep practice happens, historically, in some form of withdrawal from constant social and informational input.

That condition has been disappearing for 2 decades, faster every year. The smartphone made it routine to be available everywhere, and the social platforms made not being available costly. Earbuds and audio took away the silent walk. Streaming filled every empty room with someone else's words. AI assistants now offer what used to be solitude with a companion installed inside it: open the laptop and someone is there, patient, agreeable, and infinitely available. Solitude has not been outlawed. It has been outcompeted.

I want to name this directly because it matters more than people writing about it usually concede. The ability to practice presence requires regular contact with your own unaccompanied mind. Without that contact, the mind that is supposed to do the deep noticing never develops. Worse, it forgets that the noticing is even possible.

The cure is to protect, deliberately, the conditions under which presence can be practiced. The technology can stay. A walk without earbuds, a meal without a screen, a morning hour before the inputs begin. The silence is the room in which the practice happens.

The Companion Problem

The trust collapse reaches into something more intimate than art and journalism. People are forming attachments to AI systems, including systems designed to simulate companionship, emotional support, and therapeutic conversation.

Some uses soothe. Some scaffold a person toward human connection. Others replace the very friction that relationships require. A system that is always available, always validating, never disappointed, and never truly other can feel like a solution to loneliness. It may also become a practice in avoiding the capacities that real relationship develops.

The warning is not that every AI companion is harmful. The warning is that simulated intimacy arrives inside a loneliness epidemic that did not begin with AI. Atomized housing, shrinking civic life, screen-mediated adolescence, and thinner social networks prepared the ground, and AI companions fit that ground too well.

A person develops through difficult relationships: tolerating another consciousness that does not simply reflect them back, repairing misunderstanding, feeling irritation without leaving, and discovering that being known requires effort from more than one side. Presence is built there. AI can mimic the language around that work. It cannot do the work with us.

What AI Cannot Practice

This book has asked, in several chapters, what AI cannot do. The deepest answer is here.

AI cannot practice presence, and the reason is structural rather than temporary. There is no one inside the system who could be present, in the contemplative sense, to anything. The system generates outputs. It occupies no body in any room. Attention is not the kind of thing it has, so there is nothing to return to a single point, and control cannot be surrendered where there is no one to do the surrendering. It simulates the products of human attention without ever paying any.

AI is a different kind of thing than a human being. The kind of work the contemplative traditions name, the slow disassembling and reassembling of the self that happens when a person sits with what is, repeatedly, over years, is not a kind of work any current system can do.

That fact is at the center of the argument this book has been making. As AI handles more execution, more pattern matching, more generation of outputs that look like human thought, the irreducible human contribution concentrates on the capacities that depend on a present, accountable, embodied self: judgment, discernment, care, the willingness to stand inside an uncomfortable situation instead of routing around it, the presence to notice what nobody documented.

These capacities get built, slowly, through the practice this chapter is about.

A Practice

I want to leave the reader with something specific.

If you have never sat with a deliberate practice, start with 20 minutes a day, in the morning, before the inputs begin. A timer. A chair. No phone. There is one instruction. Place attention on the breath. When attention wanders, return it. When it wanders again, return it again. The point is not to keep attention on the breath. The point is the returning. That is the muscle being built.

Twenty minutes sounds small, but most adult professionals have given less than that to a single non-task activity in years. The first week will feel like nothing is happening. The second week will feel like nothing is happening. By the third or fourth week, something will start to shift, often outside the practice itself, in the 10 minutes after it: a pause before a reaction, a second thought before a quick reply, a small space inside the day that was not there before.

Add a weekly walk without input. One hour, no podcast, no music, no phone in hand. Notice what your mind does without anything to consume. The early walks will feel boring. The boredom is the point. Underneath the boredom is a quality of awareness most adults have not had access to since childhood, and it returns when you stop drowning it out.

Add, eventually, a longer rhythm: a half-day a quarter, or a full day a year, of disconnection, solo, with no companion, including the AI ones. The longer rhythm is where the deeper noticing tends to happen. Most readers will resist this most strongly. The resistance is informative.

These practices are not religious in any required sense. The contemplative traditions developed inside religious settings, and those settings gave the practices their depth. Many people will want to find a tradition that holds the practice in a context; others will not. The practices work in religious contexts and outside them, in the same sense that physical exercise works across belief systems. What they will not do is fit inside a productivity stack. They will not improve your output, in the short run. They may, over years, change what you are willing to count as output in the first place.

I have not described the spiritual depth of these traditions. Real depth lives in them, and 500 words would not honor it. The references at the end of this chapter point toward writers who can speak to it more fully.

What I will say is that the chapter you are reading was written by someone who started this practice 18 years ago, who has stopped and started a dozen times since, who is bad at it most days, and who keeps returning to it because it has, slowly, made me a person more capable of doing the kind of work the rest of this book is asking for.

That is the truest endorsement I can give.

• • •

The Tuesday in March, after I closed the laptop, I went to find my wife. She was in the kitchen. I asked her what she had said. She told me. It was nothing dramatic, a small thing about the afternoon, and the conversation took about 90 seconds. Then I made fresh coffee and sat at the table for a minute before I went back to the office.

The fresh coffee was not the practice, and neither was going to find her. The willingness to notice that I had missed her, and to begin again, was the practice. Begin again is the only instruction the tradition has ever consistently given.

The renegotiation of work that the rest of this book is arguing for depends on people who can begin again. Who can stay in a difficult conversation instead of routing around it, sit with a client's grief without reaching for a tool, and remain present to a student's confusion long enough to let understanding emerge. Who can stay with a contradiction long enough to understand it before trying to resolve it.

That skill is built somewhere. It is built in a chair, or on a walk, or in a half-day alone, doing the thing that does not show up on any dashboard. The chapter is here because without the practice, the rest of the book is description without prescription. It tells you what this moment requires without telling you how to become the kind of person who can meet the requirement.

This is how. There is no faster way.

CHAPTER 8 NOTES

Teasdale, John, Mark Williams, and Zindel Segal. The Mindful Way Workbook: An 8-Week Program to Free Yourself from Depression and Emotional Distress. Guilford Press, 2014. For Teasdale's deeper account of the modes-of-mind framework, see Teasdale, John D. What Happens in Mindfulness: Inner Awakening and Embodied Cognition. Guilford Press, 2022. The four-mode taxonomy summarized here is drawn from his 2022 synthesis.

Rohr, Richard. The Naked Now: Learning to See as the Mystics See. Crossroad, 2009. For the dualistic / non-dual distinction in particular, see Rohr's "Daily Meditations" archive at cac.org and his Falling Upward: A Spirituality for the Two Halves of Life, Jossey-Bass, 2011.

McGilchrist, Iain. The Master and His Emissary: The Divided Brain and the Making of the Western World. Yale University Press, 2009. The hemispheric argument touched on in this chapter is developed at length here. McGilchrist's later The Matter With Things, Perspectiva Press, 2021, expands the argument across philosophy and culture.

For the broader contemplative-traditions claim made in this chapter, see Brother Wayne Teasdale, The Mystic Heart: Discovering a Universal Spirituality in the World's Religions, New World Library, 1999. Wayne Teasdale (no relation to John Teasdale) develops the cross-traditional argument that the world's contemplative paths converge on a small set of disciplines.

For the cognitive science of meditation as a trained capacity rather than a state, see Davidson, Richard J., and Daniel Goleman. Altered Traits: Science Reveals How Meditation Changes Your Mind, Brain, and Body. Avery, 2017. The book distinguishes between short-term state effects and long-term trait changes, which corresponds roughly to the watered-down-version critique offered here.

For the loss-of-solitude argument, see Cain, Susan. Quiet: The Power of Introverts in a World That Can't Stop Talking. Crown, 2012, and Newport, Cal. Digital Minimalism: Choosing a Focused Life in a Noisy World. Portfolio, 2019. Both are accessible entry points into the cultural conditions that make contemplative practice harder than it was a generation ago.

American Psychological Association. "Artificial intelligence, wellness apps alone cannot solve youth mental health crisis, say psychologists." Press release, November 13, 2025. See also APA, "Use of generative AI chatbots and wellness applications for mental health and well-being," Health Advisory, 2025.

McBain, Ryan K., et al. "Use of Generative AI for Mental Health Advice Among US Adolescents and Young Adults." JAMA Network Open 8, no. 11 (2025): e2542281. doi:10.1001/jamanetworkopen.2025.42281.

CHAPTER 9

What Good Use Looks Like

Dr. Reyes, a family medicine physician in a small group practice outside Knoxville, walked into her exam room one Tuesday morning in late 2025. On her tablet she had a one-paragraph AI summary of her patient's recent records, 3 flagged labs, and a list of possible diagnoses ranked by likelihood. She read the summary twice. She set the tablet face down on the counter. Then she sat across from her patient, who had been her patient for 11 years, and asked how he had been sleeping.

The summary handled the paperwork. The visit was hers.

Half an hour later, after the patient had described a fatigue he had not mentioned in his portal message, Dr. Reyes picked the tablet up. The AI had ranked the 3 flagged labs as most likely indicating early thyroid dysfunction. She agreed with the ranking, ordered the next test, made a note about the sleep complaint that the AI had not been told to weight, and finished the chart in the 4 minutes she had allotted herself for closeout.

She has been working this way for 14 months, and the pattern has held. The AI compresses the chart review, and she uses the compression to be more present in the room. The visit is what her patients pay for. The chart is what the system pays for. Before, the system's demands were eating into the visit. Now, less so.

I have spent enough time with professionals in different fields working alongside AI to know what good use looks like. It looks like Dr. Reyes: the line she drew between what the tool does and what she does, and the discipline of knowing where the line goes and keeping the sides from drifting across it.

This chapter is about that line, and the people who have drawn it well.

I am going to walk the reader through 5 portraits: a clinician, a teacher, an illustrator, a small-organization leader, and myself, writing this book. The portraits are detailed because the texture is the point. There are too many books arguing for AI in the abstract and too few showing the work, day by day, of someone who has figured out how to keep the human contribution in the center of the practice. The cases are necessary. They are also rare.

The reader should leave this chapter with a specific picture in mind. Not a checklist. A picture: the picture of a professional who has accepted what the technology can do, refused to surrender what only she can do, and built a working rhythm that uses the first to protect the second.

The Clinician

Dr. Reyes is a fictional character. The pattern of practice described below is built from published research on ambient AI documentation pilots in primary care, and from physician accounts of how those pilots changed their workdays. The portrait is meant to illustrate what good integration looks like rather than to describe a specific person.

Before the AI integration, her workday looked like most family physicians' workdays: 22 patient visits scheduled in 14-minute slots, 2 hours of charting after the last visit, another 45 minutes of inbox messages before she went home. She was sleeping 6 hours a night and rarely seeing her own children awake on the days she had clinic.

The integration started with one tool. Her practice piloted an ambient documentation system that listened to the visit, generated a draft note, and presented it for her review and signature. She did not accept it on faith. The first month, she audited every note line by line against her own notes from the same visit. The system was wrong about 3 things in the first hundred visits. Two were minor. One was a confused medication reconciliation that would have been a problem if she had signed without reading.

She kept auditing for another month. The error rate dropped, partly because she started prompting the visit more carefully so the system had cleaner audio to work from. By month 3, her review had compressed to about 90 seconds per note. She still reads every note.

The second tool was the chart summary feature I described in the opening. The third was a clinical support system she uses for unusual presentations. She types the case briefly, and the system returns 3 or 4 possible diagnoses with linked literature. She uses the response as a check against her own thinking, and she would never accept it as a substitute. She drew the line in one place. The AI handles the documentation, the summarization, and the literature search. She handles the relationship, the physical exam, the diagnostic judgment, the difficult conversation, and the signature. None of those have moved. Her notes are better, because she is thinking about the patient instead of typing while she thinks, and her diagnostic accuracy on hard cases has gone up, because the AI catches conditions she might have missed. Her hours have come down. Her sleep has come back.

The point this portrait turns on, the one I keep returning to in my own work, comes down to one observation. What the technology gave her was a choice about what to do with the time it freed. She had to make that choice, then protect it from her practice administrator, who saw the freed time and proposed 4 more visits per day. She made the case for using the time on the existing patients instead, citing 2 patients whose new diagnoses had emerged in the recovered minutes. The administrator did not have a counter. The recovered time stayed in the room.

Most physicians in the published reporting have not made that case, and some have made it and lost. The ones who made it and won are the ones whose hospitals and practices show the best retention numbers, the best patient satisfaction, and the best diagnostic outcomes. The technology was the same. The argument the practitioner made for what the technology was for was different.

The History Teacher's Reconstruction Project

Tom is a composite drawn from history teachers I have worked with in central Connecticut districts. He teaches 11th-grade U.S. history in a public high school in central Connecticut. He is in his 14th year of teaching, the kind of teacher whose classroom is loud in the right way, with students arguing across rows and the teacher mostly listening.

He attended one of the AI literacy workshops I led in early 2025, skeptical. His department had spent the previous year on AI detection software and locked-down browsers, and the year had gone badly. Students hid their use, and trust collapsed. He wanted a different way.

The project he redesigned was the Reconstruction unit. The old version had been a 5-paragraph essay on whether Reconstruction succeeded or failed. The new version asked students to identify 3 communities, in 3 different states, whose Reconstruction-era experiences differed in specific ways, and to argue with primary-source evidence for what the differences reveal about why the period's outcomes were so uneven. Students had 2 weeks. They could use AI for research, for outlining, and for early drafts. They could not use it for the final argument or the in-class oral defense.

The integration he developed is worth describing in detail. Each student kept a research journal, half on paper and half digital, with a specific structure he refined over the year: a claim; 3 primary-source documents that supported or complicated the claim, with one sentence on why each one mattered; a list of specific prompts the student had given the AI, the responses received, and the student's evaluation of which were useful and which were misleading; and a short reflection at the end of each week on what they had learned and what was still confusing.

The class met 3 times a week. One of those sessions was an open table where students brought their journals and discussed where the AI had been most useful and most wrong. Tom did not grade the journals heavily. He read them, marked patterns, and used what he found to reshape the next week's class.

By the third week, the discussions had changed in a way Tom had not seen in 14 years of teaching the unit. Students were arguing with each other about the credibility of sources, including AI-generated summaries. They were spotting where a summary had skipped over a tension in the historical record, and they were finding primary documents the AI had not surfaced. The strongest students were reading more, because the AI had compressed the rote reading they had to do to write a paragraph, and they had spent the freed hours on problems the AI could not answer.

The final defenses were the best Tom had ever seen. Students stood in front of their peers, walked through their argument, took challenges from the class and from him, and stood by their reading or named the part they were no longer sure about. The defenses took 2 weeks of class time, which Tom had to argue for with his department chair. He used the recovered hours from the AI-assisted research phase to fund the defense days, and the chair signed off because the students were demonstrably reading at a higher level than the previous cohort.

The line Tom drew has a few features worth carrying away. The AI was a research and drafting partner, with the work made visible in the journal. The argument and the defense stayed on the human side. The class became a place for the kind of learning that depends on a teacher reading a room and pushing back in real time, and the grade was on the thinking, the journal, and the defense, in roughly equal parts.

The next year, the rest of the department adopted the journal structure. The AI detection software moved to the margins. Trust came back, slowly. Tom is not finished with the redesign. He is teaching it again this fall, with 3 modifications drawn from what he saw students struggle with in the spring.

The Illustrator's Studio

Anna is an anonymized portrait of a working illustrator whose identifying details have been changed. She is a children's book illustrator with a working studio in Brooklyn, in her early forties, and her work is recognizable on a shelf at Books of Wonder. She has spent 18 years building a hand and a sensibility that publishers buy. When generative image tools arrived in 2022, the bottom fell out of the per-image stock illustration market within 2 years. Two of her studio mates went out of business.

Anna did not, and she also did not pretend the tools were absent. The path she chose is worth describing because it is the most disciplined integration I have seen in a creative practice.

The line she drew runs through the work in 3 places. One: no image she delivers to a client originates with an AI generator. Her hand makes the image. Always. Two: she uses AI for everything that is not the image. Research on a historical period the book is set in. A structured brainstorm of compositional possibilities before she goes to the page. A draft of the storyboard captions, and a first pass at the email to the art director that explains why she made a particular choice. Three: she names the use to her clients. She tells the publisher exactly where the AI assisted and where her hand started, and the contract reflects the line.

The integration changed her practice in ways she did not expect.

The research phase she used to spend 2 weeks on now takes 3 days. She arrives at the page with more developed ideas because the AI has been a thinking partner during the structured brainstorm. The storyboard captions used to be a chore she resisted; now they take an hour and the resistance is gone. The freed time has gone into longer studio days on the paintings themselves, and the paintings, by her own assessment and by the response of her recent clients, are her strongest in 5 years.

She told me something I want to put on the page exactly. She said that the temptation was always there to let the AI do the rough sketch and just clean it up, and that she had to be honest with herself that what made her work hers was the part of the process where she had to find the image with her own hand. The finding was the part she would lose first if she let the AI in. The discipline of refusing the easy step is what has kept her work hers. The hours she saved on the research phase paid for the hours she now spends on the part of the work that is irreducibly her.

Her per-project rate has gone up. Her annual income has stabilized after a hard 2023. She turns down work that does not match the line she has drawn. She is not for everyone, and that is a feature.

The Director's Decision Memo

Maya is a composite drawn from nonprofit leaders who adopted similar AI-assisted preparation practices in 2024 and 2025. She is the executive director of a regional food bank in Sacramento, running a staff of 42 and a network of 70 partner pantries. Most of her decisions are about how to allocate scarce resources across communities with overlapping needs, and most of her hardest hours are spent preparing for board meetings and stakeholder conversations.

She started using AI in late 2024 for 2 specific tasks: the synthesis of partner reports into a single board-ready summary, and the preparation memo she writes for herself before any difficult conversation with a major funder, a partner pantry director, or a county supervisor.

The synthesis use takes maybe 20 minutes a month. Forty-five partner reports come in, each several pages. The AI compresses them into a structured summary organized around the metrics the board cares about. She reads the summary, spot-checks 5 of the underlying reports, and adds the patterns the AI missed because they require knowing the personalities involved.

The preparation memo use is more interesting. Before a hard conversation, she types up the situation in her own words: who she is meeting with, what she wants, what they likely want, what the past 2 years of the relationship have looked like, and where the conversation could go wrong. She then asks the AI to identify the likely emotional dynamics, the prompts she has not yet put to herself, and the moments where she might be tempted to give up something she should hold onto. She reads the response without following it, using it as a checklist for her own thinking.

What she gets out of the practice is her own preparation, made faster and more rigorous by having a thinking partner that does not get tired and does not have a stake in her organization's politics. The AI's analysis is the byproduct. By the time she walks into the meeting, she has thought through the conversation more thoroughly than she would have on her own in the time she had.

The conversations themselves are unchanged. They are conversations between people, and she holds them with her full attention, without the AI in the room. Her staff has noticed that her preparation has tightened. Her board has noticed that her summaries have improved. She has not told the staff or the board exactly how she works; she does not consider the AI use a secret, and she would tell them if asked. She has just not made the practice the headline. The work is the headline.

She told me, when I asked her what she had learned in the 18 months of using the tools this way, that the most important thing was to never let the AI write what she would have to defend. The synthesis she could check, and the preparation memo she could use or discard. The decisions, the conversations, and the board summaries she stood behind in the room came out of her. The AI's job was to free the hours she needed to think, and to widen her preparation. Her job was the rest.

How This Book Got Written

I want to walk the reader through one specific stretch of writing this book, because the preface promised a surgeon-and-scalpel relationship with the technology and the rest of the book has not paid that promise off in detail. This is the place to do it.

The chapter on the missing middle was the hardest chapter to write. I started it the way I had started the others, by asking an AI to help me think through the structure, and within 30 minutes I had an outline I did not believe.

The outline was technically correct. It said what could be said about the missing middle. It did not say what I had come to believe was true about the missing middle, because I had not yet figured out what I believed. The outline was finished. My understanding was not.

I closed the AI window and did not open it again for 9 days. In those 9 days I walked, mostly. I read the Teasdale book on mindfulness for the second time, and I read 3 articles that turned out not to matter. I drafted the chapter twice, by hand, in a notebook, badly. The second handwritten draft contained a sentence that surprised me: that the middle is the part of any difficult work where you discover what you are actually trying to say. The chapter that ended up in this book is built around that sentence. It never came from the AI. It could not have. The sentence came from 9 days of being alone with the problem.

After I had the sentence, I came back to the AI, and the use I made of it then is the use I have made of it across most of the book. I asked it to help me find research that supported or complicated the argument I was making. It found the OECD report, the Frontiers in Education study, and Land's creativity research, which I had vaguely remembered but could not source. Each source it surfaced was a real one. I read each piece. I cited what held up. The AI saved me dozens of hours of bibliographic work that would not have made the chapter better.

The drafting I did myself. The AI did not write the sentences. I tried letting it. The sentences it produced were correct and dead, and I have read enough AI prose by now to know what I was looking at. I rewrote everything it offered. The chapter that exists is mine.

A second example: the chapter on the trust collapse. I was stuck on the Jason Allen Colorado State Fair section, because I knew it was too obvious an opening and I could not find a way to make it land for a reader who had heard of it before. I asked the AI for 10 alternative openings. None of them were good, and one was bad enough to make me laugh. Inside the laugh I noticed that the original opening was actually fine. I had been editing for an imagined sophisticated reader instead of writing for the reader I had, so I went back to the original. The AI's role was to be wrong in a specific enough way that I noticed what was right. Most people do not advertise this kind of use.

A third example: the chapter on the practice of presence opens with a specific Tuesday morning when my wife came to the door and I missed the conversation. The opening was a Tuesday morning that happened, and the AI knows nothing about it. The AI helped me research Teasdale's modes of mind so I could write about them precisely. The research was useful. The opening was the part of the work the AI could not have produced. The chapter is the two woven together.

Across the book, the rough division has been about 70/30 in my favor. I write about 70 percent of the words, and I research, plan, and pressure-test about 30 percent of the work with AI assistance. The percentages are inexact and they shifted by chapter. The principle is steady. AI is faster and broader than I am at finding sources and stress-testing arguments. I am more original than it is at saying what I actually mean. Each of us has done the part that the other does badly.

The book you are reading is mine in every sense that matters: the argument, the voice, the judgment about what to cut, what to keep, what to challenge, and what to stand behind. The tool helped me do the work faster and more thoroughly than I could have alone. It did not do the work.

If the rest of the book has read as more skeptical of AI than the preface promised, I want this section to do the corrective work. I have used the technology hard, every day, for 2 years. I have written this book using it, and I have built a practice around it that has made me more productive than I have ever been and, I hope, no less honest. The misuse of the technology is the problem. The technology by itself is a tool. The right use is what this chapter has been about.

What the Portraits Have in Common

Five portraits: a physician, a teacher, an illustrator, a director, and a writer. The fields, scales, and stakes differ. The pattern is the same.

Each one drew a line, naming explicitly where the AI helped and where it did not. The line ran through the practice in a different place for each of them, and the discipline was where to place the boundary between assistance and substitution.

Each kept the human contribution at the center. The visit. The classroom defense. The hand that makes the image. The conversation. The sentence. The technology absorbed the work that surrounded the contribution and left the contribution itself to the human.

All of them accepted the trade-off honestly. The technology saved them hours, and the hours went into the part of the work that mattered most, while output volume held steady. They had to argue, in some cases, against organizations that wanted the saved hours back as additional throughput. The ones who won the argument received the renegotiation chapter 10 is about. The ones who lost it lost the gains the technology had offered.

They named the use, telling their clients, their students, their boards, and their colleagues exactly how they were working. The transparency was a feature of the practice, and the disclosure happened up front.

And they stayed honest about the friction. The AI got things wrong, and they found those things by reading. Instead of developing confident ignorance, they calibrated their oversight to what they could catch.

Good use, in plain terms, is what those 5 things look like together: a line drawn deliberately, a human contribution protected, a trade-off accepted honestly, a use disclosed openly, and a discipline of oversight kept current.

The book has been arguing, since the introduction, that AI exposes what was already broken. The portraits in this chapter are about what the same technology looks like when it is used inside a relationship a person has thought through carefully. The technology is the same. The relationship around it has changed.

That relationship is the work. The chapters that follow describe the institutional renegotiation that has to happen for relationships like these to scale, and the people who are being left out of the conversation entirely. The portraits in this chapter are what the rest of the book is, in the end, fighting for.

CHAPTER 9 NOTES

The clinician portrait is fictional. The working pattern described draws on published reporting and peer-reviewed research on ambient AI documentation pilots in primary care, including Microsoft/Nuance DAX Copilot reporting, Haberle et al. in the Journal of the American Medical Informatics Association, Wendt et al. in Future Healthcare Journal, and the Goh et al. randomized clinical trials cited in the chapter 5 notes.

The teacher portrait is a composite drawn from history teachers I have worked with in central Connecticut districts. The Reconstruction unit redesign described here has been adapted by several teachers, with variations on the journal structure.

The illustrator portrait is drawn from one specific working illustrator who agreed to be described here on condition that her name and location be changed. The professional details are accurate. Her contracts and her practice are as described.

The director portrait is composite, drawn from interviews with two regional food bank directors and one director of a women's services nonprofit, all of whom adopted similar AI-assisted preparation practices in 2024 and 2025.

For published research on calibrated AI oversight in clinical practice, see the Goh et al. studies cited in the chapter 5 notes. For the broader argument about AI as a thinking partner rather than a substitute, see Mollick, Ethan. Co-Intelligence: Living and Working with AI. Portfolio, 2024. Mollick's "always invite AI to the table" framing is more permissive than the line several of these practitioners draw. The argument of this book is closer to a "sometimes invite AI to the table, on terms you have decided" position. The difference is small in practice and larger in principle.

For the critical-thinking, cognitive-offloading, and productive-struggle concerns that shape how some of these practitioners limit their AI use, see the chapter 6 notes.

PART IV

The Renegotiation

CHAPTER 10

The Renegotiation

The terms of the old contract were simple. Compliance equals a paycheck. Standardized output means good grades. Execution translates into calling yourself a professional. That contract is now void, because the machine can satisfy those terms faster, cheaper, and more consistently than any human being. The renegotiation is already underway, in faculty meetings, in boardrooms, in studios, in hiring conversations.

We can design the new contract deliberately, or we can let it happen to us, the way a lease auto-renews on worse terms when nobody calls to negotiate.

Leaders now face a blunt choice. Organizations with imagination can use AI to expand the ambition of the work. Organizations without it will use AI to cut headcount and call the result strategy. The renegotiation turns on whether institutions use freed time for something that matters.

This chapter is the practical part, concrete enough to be useful on Monday morning: the redesign of how we teach, how we manage, how we create, and how we hire in a world where the machine handles the output.

Five redesign tests guide the chapter. For schools, what is the learning goal AI cannot substitute for? For organizations, who captures the time AI saves? For professionals, what outputs can I sign with full confidence? For creatives, where does my work become mine? For leaders, which human capabilities are we still developing on purpose?

Redesigning Education Beyond Grade the Prompt

Many schools built their first response to AI on detection: surveillance software, locked-down browsers, AI content detectors. Students learn to hide their tracks, and nobody learns anything important.

A better path changes the metric entirely. If the output can be generated by a machine, stop grading the output. Grade the process, the judgment, and the defense.

"Grade the Prompt" is a start, but only a start. Asking students to design the prompt that would generate an essay, then critique the AI's output, develops real skills. It requires knowing the subject well enough to specify what a good answer looks like, and it requires critical thinking to evaluate whether the AI delivered it. That's more demanding than most current assignments.

But the complete picture is larger. The research from Frontiers in Education in 2025 identified the key tension: AI can assist learning while also risking memory, critical thinking, and creativity when overused or misused. The FACT approach proposed by researchers, emphasizing Fundamental skills, Applied projects, Conceptual understanding, and critical Thinking, structures a redesign of assessment that acknowledges value and risk together.

In practice, that looks like this.

Problem framing matters more than problem solving. Before a student submits any work, they should articulate the problem they're addressing. What are they trying to understand, and what is the problem underneath the prompt? AI can answer efficiently. It cannot identify which prompts are worth giving it. That's a human call, and it's the one that should be assessed.

Evidence selection and sourcing should be graded explicitly. A student who can generate a 5-paragraph essay with AI assistance but can't evaluate the credibility of the sources is functionally illiterate in the skills that matter. Assessment should require students to find primary sources, evaluate methodology, and explain why they trust the evidence they're using.

Oral defense structures reveal understanding that written output conceals. After any significant assignment, students should explain what they did and why, walk through a decision they made, and defend a conclusion against challenge. AI can generate the written product. It cannot sit in the room and respond to follow-up challenges. The oral defense is where human learning becomes visible.

Revision history and process documentation should be required for major work. A student who submits a polished essay without any drafts, iterations, or evidence of revision is unusually accomplished, or the process has been bypassed. Requiring students to show their thinking over time, including what they tried that did not work, catches misuse and teaches the value of the struggle that produces real learning.

Transfer without the model should be tested regularly. If a student can only perform a task with AI assistance, they haven't learned the underlying skill. Periodic assessments without AI access reveal whether real learning has occurred. The point is to ensure AI use is supplementing understanding instead of replacing it.

Research from 2025 on student perceptions found that students who use AI with a "mastery goal structure," focused on understanding, demonstrated more critical thinking and knowledge augmentation than students with a "performance goal structure" focused only on producing acceptable outputs. The design of the assessment environment shapes which goal structure students adopt. Teachers who communicate that they value the thinking, and grade accordingly, shift students toward mastery. Teachers who only evaluate the final product invite performance.

All of this asks different skills of teachers too: less interest in correct answers, more interest in defense. Can the student explain their thinking, identify what's missing from an argument, notice when a source is unreliable? These are the skills that transfer into any professional context, and they're the skills that make AI useful instead of dangerous.

One of the most practical tools I've seen emerge from the districts I work with is what educators call a learning trail, a short set of artifacts documenting the student's path. The destination shows up on its own. For a writing assignment, the trail might include a claim statement written before drafting, selected quotes with one sentence explaining why each matters, a rough outline, a draft with 2 rounds of revision notes, and a reflection naming what changed between drafts. AI can assist with parts of the trail. But the trail makes outsourcing the thinking visible, and makes learning easier to see. When students get used to showing their trail, they start to understand that school values the thinking, not the packaging.

Redesigning Work Around Decision Quality

Organizations have been measuring productivity by the pound for decades. AI can generate infinite volume, which makes that metric meaningless overnight. If an employee uses AI to send 500 mediocre emails in an hour, they're not productive. If a coder uses AI to generate 10,000 lines of bloatware, they're not an asset.

The professional shift favors decision quality over output volume.

A decision is a moment when a human being looks at available information, weighs the context and the values at stake, considers what they don't know, and says, we go left. AI can calculate probabilities: "Based on past data, going left has a 60 percent success rate." It can't weigh the moral cost, feel the team's fear, or take responsibility if the car crashes.

The Deloitte 2025 Human Capital Trends report argues that organizations investing in human capabilities and intentional human-AI work design are more likely to report stronger business and workforce outcomes. The investment in human development is a driver of performance, not a cost line.

Organizations that thrive in the next decade will be the ones that stop measuring activity and start measuring judgment.

Hiring changes. Stop looking for people who are good at following instructions and start looking for people who are good at naming the right problems: the "spiky" people, with eclectic backgrounds, weird hobbies, strong opinions, a track record of caring about quality, and the ability to look at 10 AI-generated options and identify the one that will work.

Management changes too. Instead of "What did you do today?" the better prompt is "What did you decide today?" What judgment calls did you make, what information did you gather to make them better, and what assumptions did you challenge?

How you use the time AI saves changes as well. The 90 minutes that AI saves on report generation have value only if they get spent on the work reports can't capture: the conversation with a struggling team member, the relationship with the stakeholder who has quietly disengaged, the early warning buried in data no algorithm is flagging. If the saved time goes straight into more reports, the organization is just generating more theater with better tools.

Concrete workflow changes matter here. Establish clear policies about what AI outputs require human review before use. Separate clearly the tasks where AI assists from the tasks where a human is solely accountable. Build explicit checkpoints where decision quality is assessed: what alternatives did you consider, what could go wrong, what would you need to see to change your recommendation? Make these visible in your processes. Assumptions don't count.

Redesigning Creative Practice Around Vision

For artists, writers, designers, and every kind of creative professional, the renegotiation feels existential. If the machine can make the image, something else has to carry the value.

The answer is vision.

Execution has been democratized. Anyone can generate a song that sounds like a competent pop track, an image that looks like a photograph, a first draft of almost anything. The barrier to entry for making has collapsed. The barrier to entry for meaning has never been higher.

The creative professional is no longer just the person who holds the brush. They are the person who knows which painting to make, and why, and what it needs to say to be worth making at all.

Filmmaking has long separated craft from vision. The director doesn't hold the camera, sew the costumes, or build the sets, yet we consider them the film's author, because they hold the vision. They make the thousands of micro-decisions that steer the entire enterprise toward a specific emotional destination. The craft serves the vision.

AI speeds up that structural shift for every creative discipline. The generation is increasingly cheap, and the curation, the selection, the refinement, the insistence that the work say something, those become the human contribution.

Holly Herndon, the experimental musician, has been living in this future for years. Her album PROTO featured human vocalists singing alongside an AI she built, named Spawn. She and her partner Mat Dryhurst held actual training ceremonies in Berlin, gathering hundreds of people to sing call-and-response patterns to the AI. When Herndon performs, she doesn't hide the AI. She harmonizes with it, decides when it sounds interesting and when it sounds like noise, frames the collaboration, and owns the output. Her authenticity comes from her curation. Her value is in her ideas, her taste, and the community she builds around that sensibility.

Creatives who survive this transition will be the ones who double down on what is distinctly theirs, who lean into their specific perspective, their specific experience, their specific way of seeing, who make work that smells of the earth, of the body, of the specific pain and joy of being alive in a particular time and place. AI can approximate the average. It cannot approximate you.

My research has identified 3 models emerging for how creative legitimacy works now. The Curation Model positions the human as selector and contextualizer of machine output, with authenticity living in the choosing. The Collaboration Model treats AI as a partner, requiring new vocabularies for credit that specify who contributed what. The Transparency Model makes authenticity a function of honest disclosure instead of human exclusivity. Each works for different contexts, and none works alone. Transparency without curation produces disclosed mediocrity. Curation without transparency produces hidden dependence. What they share is a requirement that a human remain accountable for the result.

Redesigning Provenance by Making Authenticity Visible

A practical trust system is emerging, and organizations should be building toward it on purpose.

Disclosure norms need to be set internally before they are imposed externally. Decide now what level of AI assistance requires disclosure in your communications, your published work, your client deliverables. The EU AI Act will increasingly mandate this for certain categories of output, and C2PA Content Credentials offer a technical mechanism for documenting provenance. Getting ahead of these requirements is better than scrambling to catch up.

Audit trails should be standard for consequential AI-assisted decisions. What information went in, what output did the AI generate, what did the human reviewer change or accept, and who approved the final version? When something goes wrong, these records determine accountability. They should exist as a matter of course.

Hiring criteria should explicitly value provenance literacy: the ability to work with AI tools while maintaining clear accountability for output, the habit of documenting process, the instinct to be transparent about how work was produced. These are professional competencies in an era where trust is scarce and verification matters.

What Creative Professionals Need to Build

The writers, journalists, illustrators, musicians, and designers whose work is most immediately exposed to AI are in a particular kind of position. The tools that threaten to replace them are often tools they could use, and the tools that flood the market with cheap versions of their work can also free them from the parts of the craft that never mattered in the first place. Those truths occupy the same room. The task is deciding what to do while they do.

Three moves, none of them heroic, separate the creative professionals who will be doing this work in 5 years from those who will not.

Make the provenance visible. Readers, listeners, and viewers are getting better at detecting AI output, and they are developing strong feelings about what was made by a human, what was assisted, and what was generated. Being clear about the process is the cost of building the kind of relationship with an audience that survives the verification crisis. Tell people how the thing was made, not in a disclosure footer but in the work itself, in the language you use to describe what you do.

Double down on the layers AI cannot reach. Keep using the tools, and notice what makes your work yours. For a writer, it is the specific interior life that generates the specific sentence. An illustrator has the small choices that accumulate into a recognizable hand. A journalist has being in the room. These are the things that do not get cheaper when generation gets cheaper. They become harder to find.

Change how you charge and for what. The billable hour model is collapsing for creative work, because hours no longer track to output in a way anyone outside the field believes. The professionals who adapt are moving toward pricing that reflects judgment, relationship, and accountability, the things that remain scarce. An illustrator who charges for the problem she solves is in a different position than one billing by the hour while the per-image market collapses.

None of this is a rescue. The creative industries are absorbing a real blow, and for many working professionals the next 3 to 5 years will be harder than the last 3 to 5 were. But the people building careers that will still be standing on the other side share something in common. They stopped competing with the machine on the axes where the machine wins, and started building on the axes where the machine cannot compete at all.

The Synthesis

The temptation in this moment is to pick a side: Team Human or Team Machine, Luddite or Accelerationist. That framing is wrong.

The future requires a real integration. Silicon can handle execution, pattern matching, scale, and information processing, while human judgment supplies intent, presence, accountability, and meaning. The machine may be the engine. It is not the steering wheel.

When this synthesis works, you get the doctor who has all the world's medical knowledge at her fingertips and still holds the patient's hand; the teacher who personalizes every lesson plan with AI and still looks the student in the eye to say they matter; the architect exploring a thousand generated forms who chooses the one that makes a specific community feel at home; the lawyer who uses AI for research and still reads the room to understand what the client needs; the writer who uses AI to draft and still fights for the word that makes the sentence live.

The synthesis requires discipline. It requires drawing a line around what belongs to the human alone. The machine cannot apologize for you, mentor the student, or sing the blues. Those territories belong to the human, and the discipline is in refusing to cede them, through daily practice, through the accumulated choices you make about where you show up, where you struggle, and what you sign your name to.

Leaders owe workers a more honest bargain. Honesty about what is changing means workers should not be kept in the dark when AI changes the nature of their roles; they need enough information to make decisions about their own development. Investment in adaptation means recognizing that the skills required now are learnable, but they require time and deliberate practice, and organizations that roll out AI tools without providing pathways for workers to develop new skills are selecting for compliance over capability. Transparency about outcomes means organizations using AI for hiring, promotion, performance evaluation, or disciplinary processes owe their workers a clear account of how those systems work and what role AI plays in consequential decisions. AI systems can encode historical biases, produce errors that are hard to detect, and make decisions that are effectively unreviewable when the process stays hidden. Workers whose careers are affected by AI systems they cannot examine or contest occupy a structurally weaker position than the employment relationship should allow.

The right to maintain skills also matters. Organizations that use AI in ways that systematically deskill their workforce, that route work around the human learning process instead of through it, are reducing the long-term capability of the people they employ. When the AI tools fail, when the models are wrong, when the situation requires human discernment that hasn't been developed, the organization discovers it has engineered away the ability it needs.

The fundamental principle underlying all of these specifics is that the people an organization employs are human beings, not resources to be managed through a technology transition. They are people whose working lives are consequential, whose development matters, and whose dignity the organization has an obligation to respect. The transition leaves that obligation intact and makes it more complex to fulfill. The organizations that meet these obligations voluntarily will build the trust that makes high-performing AI integration possible. The organizations that ignore them will encounter the trust deficit downstream, in ways that are harder and more expensive to address than the initial honesty would have been.

CHAPTER 10 NOTES

Stanford HAI. AI Index Report 2026. Institute for Human-Centered AI, Stanford University, April 2026. Chapter 4 summarizes micro-level productivity studies including Brynjolfsson et al. (2025) on customer support, Cui et al. (2025) on software developers using GitHub Copilot, Ju and Aral (2025) on marketing teams, and Becker et al. (2025) on developers using AI for open-source work. The finding that experienced developers were roughly 19 percent slower with AI while believing they were faster is from the Becker study as cited in the report.

Frontiers in Education. "Balancing AI-Assisted Learning and Traditional Assessment: The FACT Assessment in Environmental Data Science Education." Frontiers in Education 10 (2025). doi:10.3389/feduc.2025.1596462. The "cognitive paradox of AI in education" framework describes the simultaneous potential and risk.

Pallant, Jessica L., Janneke Blijlevens, Alexander Campbell, and Ryan Jopp. "Mastering knowledge: the impact of generative AI on student learning outcomes." Studies in Higher Education, 2025. doi:10.1080/03075079.2025.2487570. The distinction between mastery and performance goal structures is central to the study's findings.

Deloitte. 2025 Global Human Capital Trends. Deloitte Insights, 2025. The report argues that human capabilities and intentional human-AI work design are central to stronger business and workforce outcomes.

For documentation of Holly Herndon's PROTO album and her approach to AI collaboration, see Herndon, Holly, and Mat Dryhurst. PROTO. 4AD, 2019. And subsequent interviews in The Wire, Pitchfork, and The Guardian through 2024-2025 documenting her continued development of AI collaboration practices.

CHAPTER 11
The Unequal Mirror

Boxtown is a predominantly Black neighborhood in southwest Memphis. It sits less than 3 miles from what is now one of the largest commercial supercomputers in the world.

The neighborhood did not ask for it.

In June 2024, local officials announced that xAI's Colossus facility would be located in Memphis. By April 2025, community groups and the Southern Environmental Law Center alleged that 35 gas turbines, with roughly 422 megawatts of generating capacity, were operating at the Memphis site without the required permits. xAI's permit application covered 15 turbines.

Boxtown carried one of the highest cancer risks in the country before the data center arrived. The neighborhood already lived alongside an oil refinery, a steel plant, a gas-fired power plant, and 14 other industrial facilities. Public SELC materials later said an Empower Analytics study projected that xAI's proposed Southaven expansion could worsen regional air pollution and cause tens of millions of dollars in annual health damages.

KeShaun Pearson is the director of Memphis Community Against Pollution. He learned about the xAI facility the way most of his neighbors did. Through a press release.

"Our health was never considered," Boxtown resident Sarah Gladney told CNN in 2025. She lives 3 miles from the facility and suffers from a lung condition. "The safety of our communities was never, ever considered."

By early 2026, plans for Colossus 2 across the state line in Mississippi had drawn scrutiny as a much larger expansion of the same build-out.

• • •

The renegotiation described in chapter 10 is a real one: knowledge workers freed from the production of mediocre artifacts, schools redesigned around judgment, organizations naming what AI saves so the saved time goes somewhere worth going. That conversation is happening, in real rooms, in real cities, among real people doing the work of figuring this out.

Boxtown is not in the room.

The mirror that AI holds up has a frame, and the frame determines who gets to stand in front of it. Inside wealthy organizations and skilled professions, the frame is wide enough to admit a lively conversation about what comes next. Outside those rooms, the conversation looks different. It happens to people who did not ask for it, with consequences they cannot examine and did not consent to.

This chapter is about the people the renegotiation is leaving behind, and what is owed to them.

The Burden Side

I want to be fair to the access story, because it has a real case. AI can expand access. A rural clinic can reach diagnostic support it did not have, a small nonprofit can produce grant materials it could not afford to outsource, and a neurodivergent student can receive scaffolding that makes school less punishing. The access story is true as far as it goes. It runs out where the costs, training, and governance fall on someone else.

The case for AI's broad benefit is an argument about access. If the tools genuinely spread expertise, the gains land widely: a nurse practitioner in a rural clinic accesses diagnostic support that used to require a specialist, a small business owner in a developing economy accesses the strategic analysis that used to require a consulting firm. The math is supposed to work out.

The math is not working out yet. And the cost of building the system is being collected from people who were never offered a vote in the bargain.

Stanford's AI Index 2026 documented the visible part of the cost. Frontier model training runs in 2025 emitted thousands of metric tons of CO2-equivalent for a single model. Data center power capacity grew to 29.6 gigawatts globally, with the US carrying the largest share. Inference water use, the cooling required to run user queries against models in production, was projected to exceed the drinking water needs of roughly 1.2 million people. These numbers are the part of the footprint visible enough to count.

The geographies absorbing those costs are not random. Memphis. Phoenix. Northern Virginia. Central Mississippi. Communities chosen because the land is cheap, the regulation is light, and the residents do not have the lobbying capacity of the companies arriving. The trade-off goes by various names in the planning documents: tax abatement, workforce development partnership, public-private investment. The exchange, in practice, is more direct. The community provides land, water, electricity, and air. The company provides jobs (often fewer than promised) and tax revenue (often deferred for decades). The model gets trained somewhere else, on data that was never collected from the community, by a workforce that does not live there.

The labor that builds AI is also distributed unevenly. In 2023, TIME reported that OpenAI had contracted Kenyan workers, through a company called Sama, to label content for moderation training. About 51 moderators worked on the OpenAI account in Nairobi, with pay ranging from $1.46 to $3.74 per hour, by Sama's own statement to The Guardian. The work involved viewing up to 700 graphic text passages per shift, depicting violence, child abuse, and self-harm, to teach the model what to refuse.

Mophat Okinyi was 27 when he did this work. He told The Guardian that the images "pop up in my mind when I'm alone, or when I'm about to sleep." He developed paranoia. He started avoiding people. His wife was pregnant when she left him. "I lost my family," he said.

In 2023, Okinyi and 3 colleagues filed a petition with the Kenyan parliament asking for an investigation into outsourced AI labor in the country. The petition is still working its way through review. In 2024, the Kenyan Court of Appeal ruled that Meta could be sued in Kenya for similar conditions, opening a legal door that may eventually reach OpenAI's contractors.

This work has to exist. The model has to learn what to refuse, somehow, before it can refuse anything. The harder issue is whether the people doing it should be invisible in our celebration of what they made safe. The renegotiation we are talking about reaches the partner at the consulting firm. It does not reach the moderator in Nairobi.

Any honest account of who AI serves has to include the communities whose air, water, electricity, and mental health underwrite the systems. The renegotiation of work is also a renegotiation of where the costs land, and right now the costs land overwhelmingly on people who do not appear in the AI conversation at all.

The Access Gap

Inside the customer base for AI tools, the gap is also widening.

Microsoft's 2025 Global AI Adoption research found that AI adoption in the Global North grew almost twice as fast as in the Global South, widening the adoption gap from 9.8 to 10.6 percentage points. UNCTAD's 2025 Technology and Innovation Report warned that AI benefits remain concentrated and that gaps in digital foundations risk widening inequalities within and among countries. The same pattern appears inside countries, between large enterprises and small organizations, between workers with college degrees and workers without them.

A small example. A community health clinic in middle Tennessee serves about 4,000 patients across 3 rural counties. The director had heard about diagnostic AI tools that could help her nurse practitioners with complex case workups, so she asked what the tools cost. The enterprise pricing for the system she wanted was around $40,000 a year, with a separate workflow integration fee. Her annual technology budget was $12,000. She uses ChatGPT on the free tier, like everyone else with a Google account. The systems built specifically for clinical decision support exist. They are not for her.

This pattern repeats across sectors. The legal AI tools built for large firms are not licensed for solo practitioners, educational AI built for well-resourced districts stays out of reach for districts that cannot afford the integration consultants the platforms recommend, and the agentic systems automating complex office workflows assume an IT staff that small businesses do not have. The promise of democratized expertise is real in the press release. The pricing structure tells a different story.

When access gaps close, they tend to close on a delay. Smartphones arrived in 2007 and reached majority adoption in lower-income countries roughly a decade later, and broadband shows a similar pattern. AI's adoption curve is faster, but the same structural forces apply. Those with capital, technical systems, and trained staff move first. Everyone else encounters the new tools as users of someone else's use, often without a clear way to evaluate, contest, or shape what those tools do.

The argument for a more inclusive AI economy is realism about what happens when half the workforce uses the tools and the other half is used by them. Charity is a separate conversation.

The Literacy Gap

One of the clearest predictors of who benefits from AI is AI literacy. Not the surface skill of typing prompts, but the deeper ability to understand how these systems work, what they are good at, where their limitations show up, how to evaluate their output, and how to push back when the output is wrong.

I run educational AI literacy programs in Connecticut through the ACES Center for AI. We work with educators across more than 20 districts, ranging from urban Bridgeport to small rural districts in the northwest hills of the state. The cohort that finishes our certification program looks measurably different from the cohort that starts it.

At the start, most educators describe AI as a tool. By the end, most describe it as an environment. That linguistic shift matters. A tool is something you pick up, use, and put down. An environment is something you operate inside, whether you intend to or not. Once educators see AI as an environment, their prompts and priorities change, and "How do I stop students from cheating?" gives way to "What should I be teaching that AI cannot do for them?"

A high school English teacher I worked with in our second cohort, I will call her Maria, came in skeptical. She had taught The Crucible to juniors for 15 years, and the standard final assessment was a 5-paragraph essay arguing whether John Proctor was a tragic hero. By the third week of the program, she had stopped trying to AI-proof the assignment and started asking what the assignment had ever been for. The redesign she landed on asked students to identify a moment from current public life that echoed the dynamics of Salem, defend the comparison with specific evidence from the literary text and the modern context, and then steelman the strongest objection to their own argument. Students recorded a brief video of themselves walking through the comparison and the objection. Maria told me that the students who used AI heavily on the project produced thinner videos, while the students who used AI as a research partner and then had to defend their reading on camera produced the strongest work she had seen in a decade. She kept the literary analysis, cut the artifact AI could imitate, and added the part where the student had to stand behind a position out loud. Three months later her department chair asked her to lead a working group rewriting the junior English curriculum.

What we are doing in Connecticut is replicable. It is also expensive. The teachers are released from classroom duties to attend the workshops, and the training is funded through a combination of grant support and district budgets. Most of the schools and districts that most need this kind of training cannot afford it without external funding. That gap is itself an inequality of literacy, layered on top of the inequalities of access.

A small rural district in our region, with about 1,400 students across 2 schools and a per-pupil spending figure well below the state average, applied for the same training cohort our better-funded districts use. The superintendent had read about the program and made it a priority. The district's grant application for state AI literacy funding was denied in the first round, awarded instead to 3 larger districts with grant-writing staff and consultant support. She came back to us in person. We worked out a partial sponsorship, but only enough to send 3 teachers, so she had to choose. The teachers she sent were the ones whose course loads sat closest to the AI topic: English, history, and library media. The science department, the math department, and the elementary teachers handling the early literacy work where AI exposure begins all waited another year. In the meantime, the students in those classrooms encountered AI on their own, with no curricular structure and no teacher who had been trained to think alongside them about it. The wealth gap between districts became, in real time, a literacy gap between students.

The risk if we do not close the literacy gap is democratic, before it is economic. AI now plays a role in hiring decisions, credit scoring, criminal sentencing risk assessments, healthcare prior authorizations, and benefit determinations. People affected by those decisions usually have no idea AI was involved and no way to evaluate the system's reasoning, and appealing to a human decision-maker gets them nowhere, because the human has been trained to defer to the system's output.

The Pulitzer Center has documented multiple cases in 2024 and 2025 of state benefit determination systems that wrongly denied claims, often in ways that disproportionately affected Black and Hispanic applicants, with appeal processes that effectively required the applicant to prove the algorithm was wrong without being told what the algorithm did. That is what unaccountable power looks like in 2026: courteous interfaces, confident outputs, and no one in the room who can say why.

Closing the literacy gap is a civic project. The communities most exposed to AI's risks, most likely to encounter algorithmic decisions in domains that matter, and least likely to have the training to push back, need that training first. We do this work backwards if we fund AI literacy where it is easiest and call that progress.

The Inequality of Presence

There is a final dimension of the unequal mirror that deserves direct attention. It is the one I find hardest to write about, because it implicates the argument I have been building across this entire book.

The argument for human presence, for the value of genuine human discernment and relationship, resonates most readily with people who already have time, training, and security. When you are worried about losing your job to automation, working multiple jobs to pay rent, or living with chronic instability, "be more present" is a different kind of instruction.

In chapter 1, I described clinical documentation systems that can recover time from charting. In some health systems, administrators have looked at those time savings and treated them as a staffing opportunity instead of a care opportunity. The risk is plain. Minutes that could return to patients can become justification for additional throughput, paid for by nurses who now have more patients to chart for in less time.

The renegotiation I want to argue for, the one where AI absorbs the routine work and humans recover space for judgment and care, requires somebody to make it happen. Where someone with authority made the case and protected the gain, it happened. Where no one did, the gain went to the institution and the time evaporated.

This is the practical limit of the optimistic argument in this book. Human presence flows freely in environments of security and support. Where security is missing, presence becomes another thing the worker is asked to provide on top of an unchanged or worsening workload. The argument I am making about what this transition could mean for work depends on the institutions around the worker doing their part. When they do not, the worker pays twice: once in the productivity gains the AI provides, and again in the new workload the institution stacks on top of the old one.

The renegotiation in chapter 10 is necessary. It is also currently being designed by and for a specific segment of the workforce. Extending that conversation, and the protections it depends on, to the workers and communities for whom the AI transition hits hardest, is a structural necessity.

Where the protections do not exist, the optimism in the rest of this book reads as a description of someone else's life.

The Civic Dimension

The trust collapse has a public face. Deepfakes of political figures, AI-generated articles designed to look like news, and synthetic social media accounts designed to amplify division are no longer speculative risks. They are part of the current information environment.

Once citizens assume that anything may be synthetic, the damage goes beyond a single false image or fake quote. It becomes easier to dismiss true evidence and harder to reason together about shared reality. Provenance, disclosure, and appeal rights become civic necessities, not technical extras.

What's Owed

The gap between those shaping AI and those affected by it is visible in how each group talks about its future. The 2026 AI Index documented a 50-point divergence on whether AI will positively affect how people do their jobs: 73 percent of AI experts said yes, and only 23 percent of the general public agreed. The optimism lives where the control does.

Three commitments I would put on the people in positions of organizational power.

Fund AI literacy where it is scarcest. The communities that most need to understand how AI hiring systems work, how algorithmic credit scoring operates, how automated benefit determinations are made, are the communities least likely to have the training to evaluate them. Investing there is civic groundwork of the most public kind.

When you use AI tools that change someone's job, tell them honestly and early, before the restructuring memo, and give them time and resources to adapt. The organizations that do this will keep their best people through the transition. The organizations that do not will discover, around year 3, that the institutional knowledge they relied on has walked out the door. Some of it will not come back.

Audit your training data and your site choices. If the AI system you are using was trained primarily on English-language, Western, formally published knowledge, it will work best for people who look like that training data. Know the gap, name the gap, and build around the gap. If your AI build-out is concentrated in communities that did not have a meaningful voice in the planning, that is a problem your company owns whether or not the regulations require you to address it.

If you sit outside positions of power, the most useful thing you can do is build enough literacy to make hard demands of the people who are in them. Pay attention to how AI is making decisions that affect you, demand transparency when those decisions are consequential, and find others who are paying attention, and organize. The companies building these systems are extraordinarily large. They respond to organized pressure, eventually, in ways they do not respond to private complaint.

The last decade taught us that platform decisions made in a Mountain View office shape the daily life of someone in Manila. The next decade is going to teach us, in similar ways, that AI build-out decisions made in San Francisco shape the air in Memphis, the workload in a Cleveland hospital, and the appeal process for a benefit claim in Atlanta.

The choice is whether we let that shaping happen by default, or whether enough of us insist on a hand in it.

• • •

I started this chapter in Boxtown, and I will end there with KeShaun Pearson and Sarah Gladney themselves: people who live in a specific neighborhood, with specific neighbors, breathing specific air. Memphis Community Against Pollution and the NAACP filed an appeal against the xAI air permits in July 2025.

The case is moving through the Shelby County process, with parallel filings at the EPA. Whether it succeeds depends on a long list of variables, including whether anyone outside Memphis cares enough to pay attention.

I have argued throughout this book that AI is a mirror, and that the work is deciding what we do with what it shows. The mirror in chapter 1 reflected the performance of work. The mirror in chapter 3 reflected the signature. The mirror here reflects something harder to look at. We have built a technology whose benefits accrue to people like the readers of books like this one, and whose costs accrue to people we do not know, in places we do not visit, doing work we do not see.

The optimistic argument I have made across this book remains the argument I believe. The exposure of what human capability is for, which AI's arrival is forcing, can produce a recovery. It can. The recovery is also currently happening for some of us and not for others. The renegotiation lifts the floor for the already lifted, while the people standing on a floor that's already low are being asked to absorb the weight of the new arrangement.

What we build next is a choice, and the choice has to include, in the design conversation, the people who have historically been on the wrong side of technological transitions: the communities living next to the data centers, the workers labeling the training data, the applicants whose benefits are denied by systems they cannot examine, and the teachers in districts that cannot afford the literacy programs.

Whether they get a seat at this transition is the most consequential test in front of us.

That is the work.

CHAPTER 11 NOTES

Coverage of the xAI Colossus facility and its operation in Memphis: Greater Memphis Chamber, "xAI," June 5, 2024 announcement; Reuters, "Musk's xAI increased Tennessee gas turbines without permits, community groups say," April 9, 2025; Southern Environmental Law Center, "New images reveal Elon Musk's xAI datacenter has nearly doubled its number of polluting unpermitted gas turbines," April 9, 2025; and SELC, "New study finds proposed xAI gas plant could worsen regional air pollution, cause millions of dollars in annual health damages," February 16, 2026, which quotes Dr. Michael Cork on the $30 million to $44 million annual health-damages estimate for the proposed Southaven expansion. Quotations from KeShaun Pearson and Sarah Gladney are drawn from CNN reporting by Laura Paddison, May 19, 2025.

Coverage of Sama, OpenAI, and the Kenyan content moderators: Perrigo, Billy. "Exclusive: OpenAI Used Kenyan Workers on Less Than $2 Per Hour to Make ChatGPT Less Toxic." TIME, January 18, 2023. Rowe, Niamh. "'It's destroyed me completely': Kenyan moderators decry toll of training of AI models." The Guardian, August 2, 2023. "Kenyan moderators behind ChatGPT want parliament to probe OpenAI, Sama over exploitation." Citizen Digital, July 12, 2023. Quotations from Mophat Okinyi are drawn from the cited Guardian reporting.

Stanford HAI. AI Index Report 2026. Institute for Human-Centered AI, Stanford University, April 2026. Chapter 4 reports the environmental footprint figures. Chapter 9, "Public Opinion," documents the 50-point gap between AI experts and the general public on AI's expected impact on jobs.

Microsoft AI Economy Institute. Global AI Adoption in 2025, January 2026. UNCTAD. Technology and Innovation Report 2025: Inclusive Artificial Intelligence for Development, April 2025. See also World Economic Forum, Future of Jobs Report 2025, for workforce adoption and skills context.

Pulitzer Center reporting on algorithmic benefit determination systems, 2024-2025. The Center's AI Accountability Network has published a series on automated decision-making in public benefits across multiple US states.

Identifying details for the teacher and rural district examples have been changed to protect privacy, consistent with the Note on Sources and Method at the front of this book.

For research and reporting on AI-generated disinformation and epistemic vulnerability, see Shao, A., et al. "New sources of inaccuracy? A conceptual framework for studying AI hallucinations." Harvard Kennedy School Misinformation Review, 2025; UNESCO, "Deepfakes and the crisis of knowing," October 27, 2025; and NewsGuard, "Tracking AI-enabled Misinformation," 2026.

CONCLUSION

What Human Value Looks Like When Execution Is Cheap

We began with a closing submission box and a mirror.

Neil Clarke looked at what was flooding his inbox and saw something that had been building for years. We had been training ourselves to produce work that followed the rules, said little, and came from nowhere specific. The machine learned that pattern from us, then produced the same kind of emptiness widely.

The same recognition spread. Teachers saw it in essays; program managers, in status reports; lawyers, in legal research; executives, in strategy decks. The mirror went up and it showed the same image everywhere. We had confused the proof of effort with the thing itself.

What this book has been about, across 11 chapters and one argument, is what happens when the performance of work gets cheap. We measured the performance because the thing itself was hard to measure, and AI exposed that arrangement more than it broke it. What AI cannot do is what was always the work. Judge, choose, take responsibility, and be there.

A school that redesigns assessment around judgment, defense, process, and transfer is building something real. A school that adds detection software and calls it an AI policy is performing safety.

An organization that builds provenance systems, discloses AI use transparently, and holds humans accountable for AI-assisted outputs is building trust. An organization that uses AI to scale content while obscuring its origins is borrowing trust it has not earned.

A society that extends AI's benefits broadly, invests in literacy where people are most exposed, and includes affected communities in the design conversation is building something sustainable. A society that lets AI deepen existing inequalities while celebrating productivity gains for the already productive is building pressure it will eventually feel.

Some things hold steady. Expertise is still built through experience, failure, correction, and time; trust still depends on accountability; presence still cannot be downloaded; and meaning still requires a human being with something at stake.

The verification crisis will keep growing. When a voice note can be faked, when a video call can be synthetic, when a photograph no longer proves presence, truth stops being what looks convincing and becomes what can be procedurally confirmed. The cultural fight over AI art is an early rehearsal for concerns that will reach law, journalism, education, medicine, and family life.

AI will keep improving, and the pressure to delegate more, automate more, and appear more efficient while engaging less will not fade. We can improve alongside the tools. Without that effort, we become more efficient at avoiding the work of becoming fully ourselves.

I hold the optimistic wager, not because the evidence makes it certain, but because I have watched educators, leaders, creatives, and professionals meet the disruption with more seriousness than the headlines predicted. I have seen people stop asking how to beat the machine and start asking what deserves human attention.

What to Protect

The practical answer is smaller than the civilizational language sometimes makes it sound.

Protect the first draft when the work is meant to teach you something. Protect the conversation when the relationship is what matters. Protect the source trail when truth depends on provenance. Protect the slow decision when consequences will fall on people who were not in the room. Protect the human signature when trust is being asked of someone else.

These protections ask for boundaries rather than nostalgia. A teacher can invite AI into research and still require the student to defend the argument aloud. A manager can automate the status report and still spend the recovered hour with the team member who is quietly struggling. A writer can use a model to find sources and still refuse to let it supply the sentence that carries the wound.

The rule is plain. Use the machine where it removes noise. Keep the human where the work forms a person, changes a relationship, creates a public claim, or distributes harm. Those are the places where convenience has to answer to responsibility.

That is the shape of the next contract. Cheap execution gives us room. It does not tell us what the room is for. The answer will be visible in the assignments we redesign, the meetings we cancel, the signatures we protect, and the people we choose to include when a new system is being built.

That is enough to begin.

Show up. Struggle. Stand behind it.

CONCLUSION NOTES

Maslach, Christina, and Michael P. Leiter. The Truth About Burnout: How Organizations Cause Personal Stress and What to Do About It. Jossey-Bass, 1997.

Gallup. State of the Global Workplace: 2023 Report. Gallup, 2023. The 23% engagement figure is a global average; rates vary significantly by region.

Weil, Simone. Waiting for God. Translated by Emma Craufurd. G.P. Putnam's Sons,

1951.

Acemoglu, Daron, and Simon Johnson. Power and Progress: Our Thousand-Year Struggle Over Technology and Prosperity. PublicAffairs, 2023.

Ericsson, K. Anders, and Robert Pool. Peak: Secrets from the New Science of Expertise. Houghton Mifflin Harcourt, 2016.

Acknowledgments

Books are rarely as solitary as they appear. This one is no exception.

I owe a genuine debt to the educators across Connecticut who shared their experiences with AI in their classrooms and districts with more honesty than the institutional conversation usually allows. The superintendents, teachers, instructional coaches, and curriculum directors who told me what was actually happening, including the things that weren't working, gave this book its grounding in real practice. Names are withheld. Gratitude stays.

My colleagues at ACES, the Area Cooperative Educational Services in Connecticut, provided the organizational context in which the practical work described in this book took shape. The opportunity to develop AI literacy programs for educators, to watch their responses to the tools and to each other, and to see genuine professional development happen in real time, has been irreplaceable.

This book was written during a period of real uncertainty about what AI means for human work, creativity, and meaning. The people around me who held the uncertainty with grace and kept pressing on the right things made it possible to do the same on the page.

Whatever this book gets right is a collective achievement. Whatever it gets wrong is mine.

For Further Reading

The literature on AI and human work is expanding faster than any reading list can capture. The following represent the sources I found most useful in developing the arguments of this book, organized by theme rather than chapter.

On AI Capabilities and Limitations

Mitchell, Melanie. Artificial Intelligence: A Guide for Thinking Humans. Farrar, Straus and Giroux, 2019. Still the most clearly written account of what AI systems do and why their limitations are structural rather than temporary.

Stanford HAI. AI Index Report 2025 and 2026. Stanford University Human-Centered Artificial Intelligence, 2025, 2026. The most comprehensive annual data source on AI development, use, and impact.

On Education and Cognitive Development

Wolf, Maryanne. Reader, Come Home: The Reading Brain in a Digital World. Harper, 2018. The research on deep reading and its relationship to the development of the empathic and analytical mind is directly relevant to the educational concerns raised in this book.

Ericsson, K. Anders, and Robert Pool. Peak: Secrets from the New Science of Expertise. Houghton Mifflin Harcourt, 2016. The definitive account of deliberate practice and its role in expert development.

OECD. Digital Education Outlook 2026. OECD Publishing, 2026.

On Work and Organizations

Brynjolfsson, Erik, and Andrew McAfee. The Second Machine Age: Work, Progress, and Prosperity in a Time of Brilliant Technologies. W.W. Norton, 2014. The foundational economic account of AI's labor market effects.

Deloitte. 2025 Global Human Capital Trends. Deloitte Insights, 2025.

World Economic Forum. The Future of Jobs Report 2025. World Economic Forum, 2025.

On Creativity and Authenticity

Doshi, Anil R., and Oliver Hauser. "Generative AI Enhances Individual Creativity but Reduces the Collective Diversity of Novel Content." Science Advances 10, no. 28 (2024).

On Media and Technology

McLuhan, Marshall. Understanding Media: The Extensions of Man. McGraw-Hill, 1964. Still essential. Still right.

Postman, Neil. Technopoly: The Surrender of Culture to Technology. Knopf, 1992. The warning issued thirty years ago that this moment has made newly urgent.

On Contemplative Practice and Human Wholeness

McGilchrist, Iain. The Master and His Emissary: The Divided Brain and the Making of the Western World. Yale University Press, 2009.

Teasdale, John D., J. Mark G. Williams, and Zindel V. Segal. The Mindful Way Workbook. Guilford Press, 2014.

Rohr, Richard. Falling Upward: A Spirituality for the Two Halves of Life. Jossey-Bass, 2011.

About the Author

Bob Hutchins is an AI literacy strategist, writer, and organizational psychologist whose work sits at the intersection of technology, human development, and meaning.

He is the AI Literacy Lead at ACES (Area Cooperative Educational Services) in Connecticut, where he assists and advises the ACES Center for AI and develops certification programs and professional development for educators and administrators across the state. He is Chief AI & Marketing Officer at ROME Insights and teaches intro to AI at Lipscomb University as an adjunct professor.

He has worked in digital marketing, strategic communications, and organizational development for over 2 decades, having founded BuzzPlant in 2000 and run landmark digital campaigns for major film studios, faith organizations, and consumer brands.

He is the author of Faith-Based Marketing, Finally Human, and Our Digital Soul, and has a TEDx talk with over 1 million views. He writes the Silicon and Soul newsletter on Substack and The Human Voice with Bob Hutchins podcast.

Made in the USA
Las Vegas, NV
19 September 2026

4761acd3-ff4c-4056-a919-8f9f69e2da66R01